Spark to Mind: Brain's Epic Journey

Namita

Table of Contents

Chapter 1: Introduction

1.1 Background

The brain changes dramatically in its appearance during gestation and even after birth. These changes include enlargement of the cerebral hemispheres and expansion of the occipital, frontal, temporal and parietal lobes. The primary, secondary and tertiary sulci separate numerous folds called gyri. The convoluted nature of the brain surface greatly increases its surface area, without increasing the overall size and volume. At birth the human brain is about 25% of its adult weight (1336 grams - 1198 grams in adult males and females respectively according to Hartmann *et al.,* 1994), crossing the 50% mark at 6 months, 75% at 2 years, 95% at 6 years, and 100% at 7 years of age (Learning, 2020).

After birth brain (especially within the first two years) development is extremely dynamic and could possibly play a role in neurodevelopmental disorders, including schizophrenia and autism (Knickmeyer *et al.*, 2008).

The development of the venous system is characterized by the formation of a highly irregular network of capillaries and the expansion of certain channels into definitive veins. Due to these multi-channelled beginnings the adult venous system is characterised by a higher incidence of anatomical variations than seen in the arterial system (Carlson, Cardiovascular system, 2014).

Literature on the cerebral venous drainage system is limited and is mostly based on the work of George Linius Streeter and Herbert McLean Evans in the early 1900's as well as Dorcas Hager Padget in the 1950's. The work published by JF Knott "On the cerebral sinuses and their variations" was published in 1881 (Knott, 1881). It is acknowledged that the work made a great contribution to the understanding of the dural venous sinuses at that time. However, it was decided to quote and use more recent literature on the subject as in the work of Streeter and Padget. Subsequent investigations into the embryological development and normal anatomy of the cerebral venous system were done in later years by various authors, however there remain gaps in the literature (Butler, 1967; Bisaria, 1985; Pryd & Edwards, 1996). New investigations into the cerebral venous system and its variations need to be performed to build on and add to the literature and our understanding of dural venous drainage patterns and its embryology.

Advances in non-invasive imaging procedures allow for an increased understanding of the variations of the dural venous sinuses. The confluence of sinuses shows the most abundant variations, however a comprehensive classification including all variation types is lacking in the literature (Bayarogullari *et al.,* 2018).

1.1.1 Research statement

To describe variations in the pattern of venous drainage of the brain in humans from childhood to adulthood, to speculate on the possible mechanisms behind the development of such anatomical variations and to discuss the clinical impact in cerebrovascular imaging and surgical interventions.

1.1.2 Aims of the study

- Describe the embryological and fetal development of venous drainage patterns of the brain by means of a literature review and a limited number of dissections
- Describe the macroscopic anatomical venous drainage patterns in children (birth to 12 years), adolescents (13 – 17 years), adults (18 – 64 years) and elderly people (65 years and older) using radiological imaging
- Speculate on the possible developmental mechanisms behind variations observed in the cerebral venograms
- Discuss the clinical impact of cerebral venous drainage patterns in cerebrovascular imaging and surgical interventions and identification of persons

1.1.3 Objectives of the study

The sample population for this study was grouped according to age, grouped as children (birth to 12 years), adolescents (13 – 17 years), adults (18 – 64 years) and elderly people (65 years and older). For each of the age groups the following objectives were identified:
- Describe macroscopic anatomical venous drainage patterns using cerebral venograms obtained from the Department of Radiology at the Red Cross Children's War Memorial Hospital (RCCH) and the Department of Radiology at the Groote Schuur Hospital (GSH).
- Study the venous drainage patterns in a small number of fetuses by dissection to confirm what is described in the existing literature
- Describe the different patterns of venous drainage in adolescents, adults and elderly people
- Compare the macroscopic anatomical venous drainage pattern between age groups
- Compare the macroscopic anatomical venous drainage pattern between sexes
- Determine if there is a statistically significant change in the venous drainage pattern from one age group to the next
- Compare venous drainage patterns for the three age groups between females and males
- Determine if there is a statistically significant difference in the venous drainage pattern between females and males
- Discuss the implications of variations of the venous drainage patterns in clinical and anthropological settings
- Speculate on the relationship between the presence of radiographically visible cerebral pathology and presence of variations in the venous drainage pattern

1.2 Anatomical nomenclature

1.2.1 Herophilus

Herophilus made phenomenal anatomical observations of the human body. His observations contributed to the current understanding of the brain, eye, liver, reproductive organs and nervous system. Herophilus was the first person to perform systematic dissections of the human body and is acknowledged as the Father of Anatomy (Bay & Bay, 2010).

1.2.2 Jean Baptiste Pauline Trolard

For more than a century after his death, Jean Baptiste Pauline Trolard's name still lives on in reports in the medical literature, primarily due to his extensive work on the anastomotic veins of the cerebral circulation. More specifically the great anastomotic or superior anastomotic vein of Trolard (Loukas, *et al.,* 2010).

1.2.3 Galen of Perganum

The two internal cerebral veins join to give raise to the vein of Galen also known as the great cerebral vein. The vein is named after the Greek physician and philosopher, Galen of Perganum. Galen had a great influence on medical practice and teachings in Europe from the Middle Ages until the mid-17th century.

Figure 1.1: Galen of Perganum. From the National Library of Medicine (image ID:192922). Retrieved December 22, 2019, from https://kids.britannica.com/students/article/Galen/274474. Used with permission.

1.2.4 Charles Labbé

The vein of Labbé, otherwise known as the inferior anastomotic vein, is named after French anatomist, Charles Labbé. The vein of Labbé is a very important structure in the superficial venous drainage of the brain, (Bartels & Van Overbeeke, 1997).

1.2.5 Friedrich Christian Rosenthal

Friedrich Christian Rosenthal was a German anatomist and surgeon. Known for his work on the olfactory system and ichthyology, his later works included descriptions of the eponymous canal in the cochlea and the basal cerebral vein. The basal vein also bears his name; the basal vein of Rosenthal (Binder *et al.,* 2006).

Figure 1.2: Friedrich Christian Rosenthal. Retrieved December 22, 2019, from Used with permission. **Table 1. 1 Table showing the change in nomenclature of the veins of the cerebrum.**

Original term	Updated term	Source
Vein of Trolard	Superior anastomotic vein	Loukas, *et al.,* 2010
Vein of Labbé	Inferior anastomotic vein	Bartels & Van Overbeeke, 1997
Vein of Galen	Great cerebral vein	National Library of Medicine
Basal vein of Rosenthal	Basal cerebral vein	Binder *et al.,* 2006

A visualization of the veins of the cerebrum is shown in figure 1.3 below.

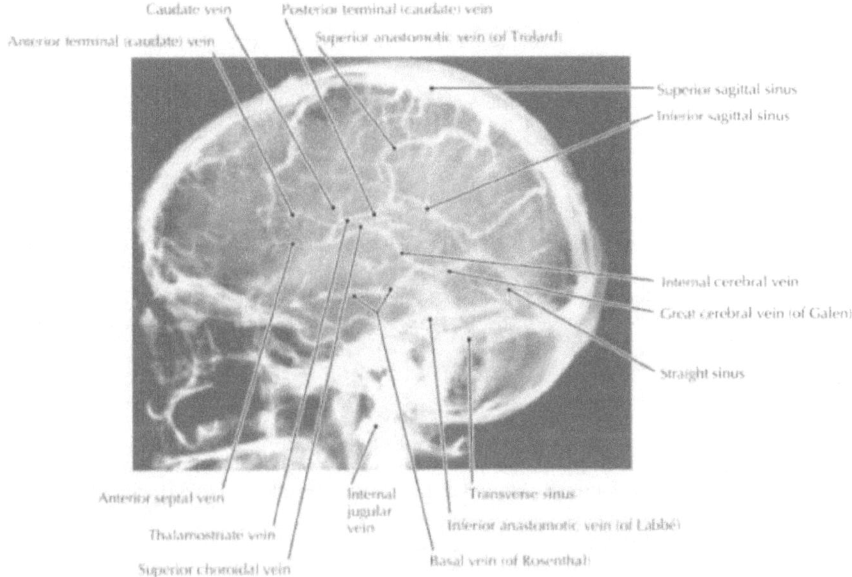

Figure 1.3: Magnetic resonance image of the veins of the cerebrum. Retrieved 04 January 2020, form Netter's Atlas of Neuroscience (pg. 120), D. L. Felten, 2016, Philadelphia: Elsevier. Copyright 2016 by Elsevier. Used with permission.

Chapter 2: Literature Review

2.1 Embryology

The human embryo starts its journey at conception. Throughout development remarkable changes occur.

2.1.1 General embryology

The human development process begins when a sperm penetrates an oocyte at the time of fertilisation. A zygote is a fertilised oocyte. The zygote is a highly specialised, totipotent cell, it can differentiate into any type of cell within the human body. Through mitosis and meiosis, the zygote will divide numerous times to gradually form a multicellular human being, going through different stages as a morula, blastocysts (in mammals), embryo and finally a fetus (Moore; 2016).

2.1.1 Age determination

The measurement made from the top of the head (crown) to the bottom of the buttocks (rump) on human embryos and fetuses is known as the crown-rump length (CRL). Crown rump length can be used to determine gestational age and an expected date of delivery. During early pregnancy it is recommended that one uses fetal crown-rump length over last menstrual period when determining gestational age (Kumar & Dubey, 2017).

Fertilisation age is used to date the human development from the day of fertilisation, which is usually 14 days after the first day of the last menstrual period. Gestational age or menstrual age; is calculated from the first day of the last menstrual period. Thus, fertilisation age is two weeks later than the gestational age (Hill, 2018).

2.2 Neurulation

At the time of neurulation, the embryo is a trilaminar disc, comprising of the ecto-, meso- and endoderm layers. The mesodermal notochord induces changes in the overlying ectoderm, causing it to become thickened forming the neural plate. The notochord is located cranial to the primitive streak. The neural crest cells are found on either side of the neural plate. The neural crest cells begin to approach each other as the neural tube is formed by the folding over of the sides of the neural plate (refer to figure 2.1). During its formation the neural tube starts to sink below the ectoderm. Some of the neural crest cells pinch off to form ganglia throughout the trunk and a variety of tissues throughout the body. This process is known as neurulation and advances both cranially and caudally from a multisite point of fusion (Gilbert, 2000).

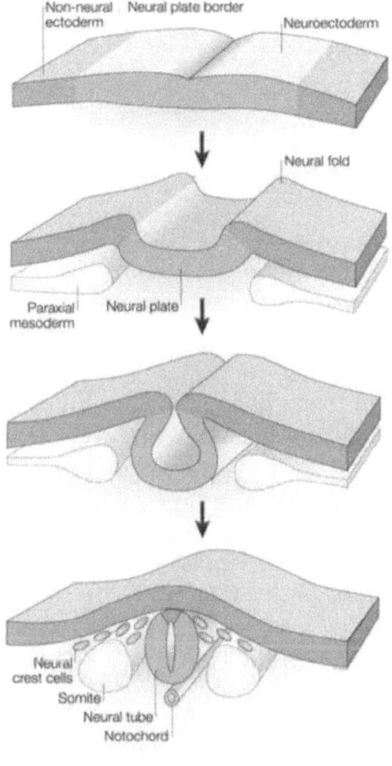

Figure 2.1: The process of neurulation. Retrieved 15 December 2018. From Neural crest specification: migrating into genomics, by Gammill, L., and Bronner-Fraser, M, *Nature Reviews Neuroscience* 4, 795–805 (2003). Used with permission.

Early development

During the early development of the embryo a forebrain (prosencephalon), midbrain (mesencephalon) and hindbrain (rhombencephalon) can be distinguished (refer to figure 2.2). The forebrain later develops into the diencephalon and telencephalon, while the hindbrain forms the myelencephalon and metencephalon. The telencephalon will form various components of the forebrain such as the future cerebral hemispheres (Dafny, 2020).

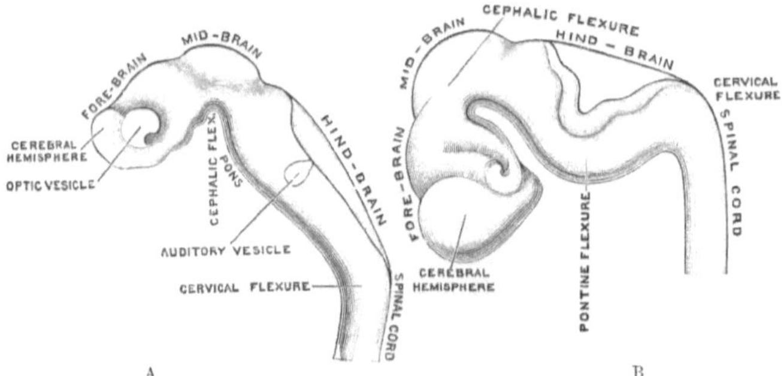

Figure 2.2: Development of the major brain structures. From Textbook of Anatomy, by D.J. Cunningham, 1903, New York, NY: William Wood and Co. Used with permission.

By the third month of gestation the major structures and regions of the brain are distinguishable. The diencephalon and mesencephalon are overgrown by the cerebral hemispheres from the telencephalon. The metencephalon differentiates into the cerebellum, while the medulla oblongata forms from the myelencephalon (Dafny, 2020).

2.3 Development of the cerebral vasculature

Around 18 days post fertilisation (3rd week of pregnancy) the vascular system starts to develop in the wall of the yolk sac. A temporary solution for the immediate needs of the embryo is yolk sac haematopoiesis, as the tissues that normally produce blood cells in an adult have not yet been formed (Carlson, 2014).

2.3.1 Arterial system

2.3.1.1 Pharyngeal (Branchial) Arches

On both sides of the foregut there are transverse swellings of mesenchyme, known as the pharyngeal (branchial) arches. Each pharyngeal arch is separated by a pharyngeal groove of ectoderm on the outside of the arch and pharyngeal pouches of endoderm on the inside. There are six pairs of pharyngeal arches in mammals, numbered one to six. These arches will form most of the structures in the face and neck, except for arch five which does not usually form any persisting structures (Robinson & Moss-Salentijn, 2020). Upon review of the work by Anderson *et al.,* (2018) as well as Gupta *et al.,* 2016, it is noted that these authors attribute malformations and adult variations to the presence of the fifth arch artery. Each arch has a cranial nerve, a rod of cartilage, skeletal muscle and an aortic arch artery associated with it (Robinson & Moss-Salentijn, 2020).

2.3.1.2 Development of the aortic arches and their derivatives

Between day twenty-two and twenty-four the first pair of aortic arch arteries forms. Due to body folding the cranial ends of the attached aortae fold into a dorsoventral loop. This results in the first pair of aortic arches lying within the thickened mesenchyme of the first pair of pharyngeal arches on both sides of the pharynx (Schoenwolf *et al.*; 2015).

The first arches will start to regress as the other arches start to form. By day twenty-six the second pair of aortic arch arteries arises and connects the aortic sac to the dorsal aortae. On day twenty-eight the third and fourth aortic arch arteries start to arise while the first arch arteries are regressing. The sixth arch arteries start to form on day twenty-nine while the second arch arteries regress (Schoenwolf *et al.*; 2015).

By day thirty-five the connections between the dorsal aortae and the third to fourth arch arteries will disappear bilaterally. Thus, the third aortic arch arteries supply blood to the cranial extensions of the dorsal aortae which in turn supply the head. Both common carotid arteries and the proximal portion of the right and left internal carotid arteries arise from the third arch arteries. The distal portion of the internal carotid arteries arise from the cranial extensions of the dorsal aortae, while the external carotid arteries will arise from the common carotid arteries (Schoenwolf *et al.*; 2015).

The connections between the right sixth aortic arch artery, right dorsal aorta and fused midline dorsal aorta are lost by the seventh week of development. However, the right dorsal aorta will remain connected to the right fourth arch artery acquiring the seventh intersegmental artery. The right subclavian artery will arise from the fourth arch artery, a segment of the right dorsal aorta and the right seventh intersegmental artery. In the region of the aortic sac connected to the right fourth arch artery the brachiocephalic artery will arise (Schoenwolf *et al.*, 2015).

The left fourth aortic arch artery will remain connected to the fused dorsal aorta and together with a segment from the aortic sac will become the arch of the aorta and the most proximal portion of the descending aorta. The remainder of the descending aorta will form from the fused dorsal aortae. The left subclavian artery arises from the left seventh intersegmental artery, which forms from the paraxial mesoderm (Schoenwolf *et al.*, 2015).

The sixth pair of aortic arch arteries arise from the proximal end of the aortic sac. The ductus arteriosus, which connects the left pulmonary artery to the descending aorta, is formed from the distal part of the left sixth aortic arch artery (Carlson, 2014; Schoenwolf *et al.*, 2015).

The external carotid arteries (supplying the face) and the internal carotid arteries (supplying the frontal part of the base of the brain and forehead) arise from the aortic arch system. At the level of the spinal cord the basilar artery forms from a fusion between the vertebral arteries growing towards the brain. The basilar artery runs along the ventral surface of the brain stem, supplying it with a series of paired arteries. As the basilar artery approaches the level of the diencephalon and the internal carotid arteries, branches from each of these major vessels grow out and fuse, thus forming the posterior communicating arteries. The basilar and carotid circulations are joined by the posterior communicating arteries. Another two branches coming from the internal carotid artery will fuse cranially in the midline to complete the vascular ring, the *circulus arteriosus cerebri*. The *circulus arteriosus cerebri* encircles the pituitary stalk and optic chiasm as it lies below the diencephalon. This arterial ring is a structural adaptation ensuring continuous blood flow to the brain in the event of occlusion of some of the major arteries supplying the brain (Carlson, 2014).

The ring-like network of arteries, the *circulus arteriosus cerebri*, consists of the six major cerebral arteries, the internal carotid arteries and the basilar artery. The ring-like structure plays a significant role in the maintenance of blood flow through the brain; with flow entering from the internal carotid arteries and the vertebrobasilar system, it enables a collateral flow between the hemispheres and provides compensatory collateral flow in case of flow disruption in one of the major supplying arteries (Mukherjee *et al.*, 2017).

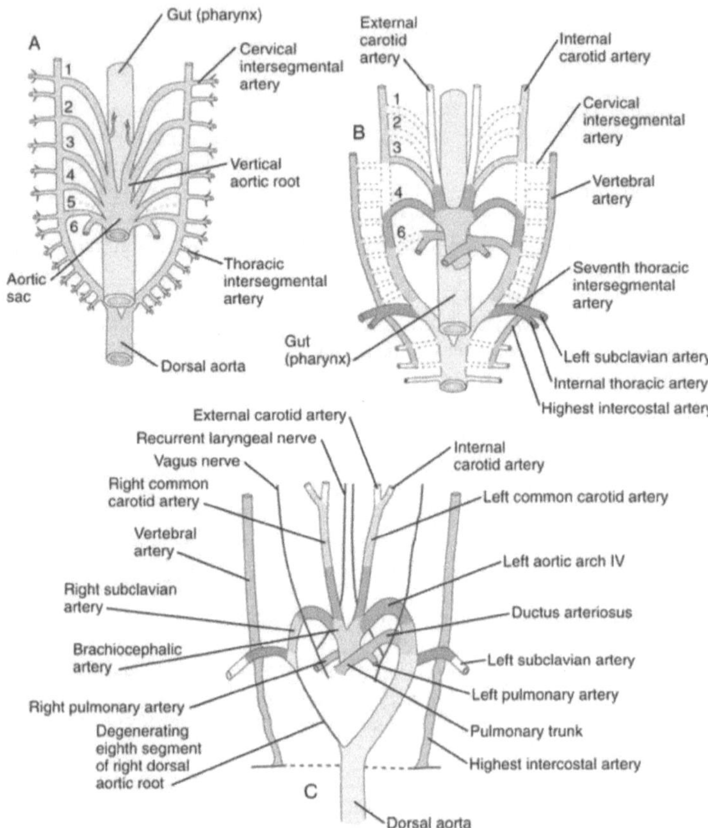

Figure 2.3: A, Schematic representation of the embryonic aortic arch system. B and C, Later steps in the transformation of the aortic arch system in a human. Disposition of the recurrent laryngeal nerve in relation to the right fourth and left sixth arch is also shown in C; by B.M Carlson, 2014, *Human Embryology and Developmental Biology* (pg. 416). Elsevier. Used with permission.

2.3.2 Venous system

Three major venous systems fulfil their function within different regions of the embryo. These systems are the vitelline system (draining gastrointestinal tract and gut derivatives), the umbilical system (carrying oxygenated blood from the placenta) and the cardinal system (draining the head, neck and body wall). Initially all three are bilaterally symmetrical and converge on the left and right sinus horns of the sinus venosus (Fasouliotis *et al.*, 2002; Schoenwolf *et al.*, 2015).

The cardinal vein system develops during the third and fourth weeks and it is bilaterally symmetrical. The cardinal system consists of paired anterior (cranial) and posterior (caudal) cardinal veins. These veins join near the heart to form short common cardinals which empty into the sinus horns. The supracardinal and subcardinal veins at first supplement and later replace the posterior cardinal veins. These two additional systems undergo extensive remodelling during development (Schoenwolf *et al.*, 2015).

The left and right subcardinal veins arise from the base of the posterior cardinals (by the end of the 6th week of gestation). They continue to grow caudally in the medial part of the dorsal body wall. These subcardinal veins connect to each other by numerous median anastomosis (by the 7th and 8th weeks of gestation) and will also form lateral anastomosis with the posterior cardinal veins. By the ninth week of gestation the longitudinal segment of the left subcardinal vein regresses and the structures on the left side of the body are served by the subcardinal system draining solely through transverse anastomotic channels to the right subcardinal vein. The original connection between the subcardinal vein and the posterior cardinal vein will be lost, making way for new anastomosis with the segment of the right vitelline vein just inferior to the heart. This anastomosis forms the portion of the inferior caval vein between the kidneys and the liver. Thus, the inferior vena cava now drains blood from the organs initially drained by the left and right subcardinal veins (Schoenwolf *et al.*, 2015).

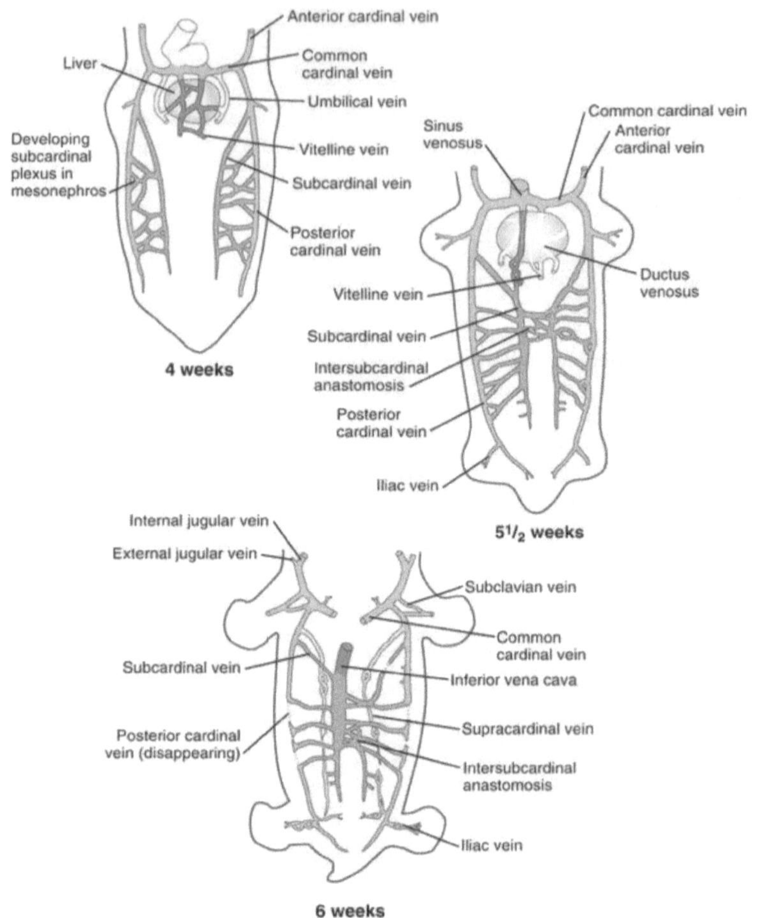

Figure 2.4: Venous development week 4 to 6 post fertilisation. By B.M Carlson, 2014, *Human Embryology and Developmental Biology* (pg. 420). Elsevier. Based on: The Development of the vena cava inferior in man, by C.F.W. McClure & E.G. Butler, 1925, *American Journal of Anatomy,* 35, pp. 331-383. Used with permission.

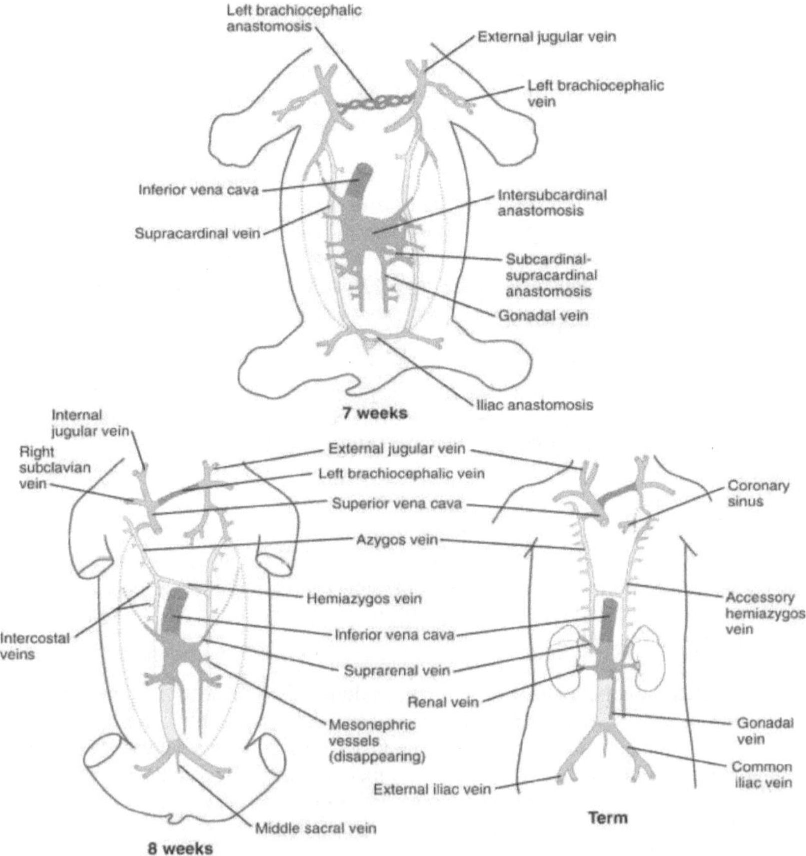

Figure 2.5: Development of the cardinal vein system 7 weeks post fertilisation to term. by B.M Carlson, 2014, *Human Embryology and Developmental Biology* (pg. 421). Elsevier. Based on: The Development of the vena cava inferior in man, by C.F.W. McClure & E.G. Butler, 1925, *American Journal of Anatomy,* 35, pp. 331-383. Used with permission.

2.4 Embryology of the dural venous sinuses

2.4.1 The dura - arachnoid interface

The dura and arachnoid layers are easily separated from each other. Separation of these two layers creates an apparent space. The apparent space that forms may be seen as an artefact of cleavage through a tissue plane (Mack *et al.*, 2009).

An investigation into the embryological origin of the dura mater shows that no apparent space exists. Initially the brain and skin is separated by a cellular network which forms from the meningeal mesenchyme (derived from the neural crest cells). This network is traditionally known as the meninx primitiva. The meninx primitiva can be divided into an inner and outer layer. The inner endomeninx and outer ectomeninx differentiate into the pia-arachnoid and dura of the neurocranium respectively. The dura consists of inner and outer layers. The inner layer which forms the dural folds (the falx cerebri, falx cerebelli and tentorium cerebelli) and it contains the dural sinuses. The outer layer forms the inner periosteum of the skull (Mack *et al.*, 2009).

The arachnoid and the dura do not separate from each other during development, thus an anatomical subdural space, comparable to the other anatomical spaces (for example the peritoneal and plural cavities) does not form. However, at the interface between the arachnoid and the dura there is a distinct soft-tissue layer. This layer contains no prominent extracellular spaces or extracellular collagen, it is however, characterized by flattened fibroblasts with sparse intercellular junctions (Mack *et al.*, 2009).

The most frequently used term to describe this interface layer is the "dural border cell layer", introduced by Nabeshima *et al.* in 1975. Thus, one could describe a collection of fluid or blood within the disrupted layers of the dural border cell layer (Mack *et al.*, 2009).

2.4.2 Bridging veins

The meninx primitiva differentiates into distinct but contiguous layers; equally the venous drainage is also differentiating into discrete systems. It is important to note that the bridging veins is formed when the venous drainage of the brain separates from the venous drainage of the dura (Mack *et al.*, 2009).

A primitive endothelial plexus of venous connections exists between the dural and pial layers. Most of the primary venous anastomoses between the pia and dura are resorbed when the primitive endothelial channels draining the brain and dura undergo venous cleavage. Definitive bridging veins are formed from a small number of venous anastomoses that remain. The superior cerebral veins collect blood from the veins draining the cerebral hemispheres. Ten to eighteen large bridging veins form from these superficial veins. These bridging veins will eventually penetrate the dura-arachnoid interface layer. These bridging veins may travel within the dura for variable distances before ultimately draining into the SSS (Mack *et al.*, 2009).

It is speculated that if these primitive endothelial channels do not undergo venous cleavage or are resorbed, it could result in the persistence of embryological venous drainage pattern; resulting in venous blood pools sub-durally in the adult.

2.4.3 Dural veins

Formation of the bridging veins occurs during the first 10 weeks post fertilisation; through the process of venous cleavage. However, the dural venous structures into which these bridging veins drain, will be modified throughout gestation. The major dural venous sinuses will only reach their final configuration postnatally. Early fetal dural venous connections are constantly changing and are plexiform in nature. The primitive dural plexus is continually adjusting due to the dramatic growth of the cerebral hemispheres antenatally. These adjustments are largely possible due to spontaneous migration of the principle dural veins, and for this a venous plexus is essential (Mack *et al.*, 2009).

A fundamental element of the embryological development of the intracranial venous system is the separation of the veins intrinsic to the brain and dura early in the development process. This cleavage process is essential to the establishment of the blood-brain barrier. The process of venous cleavage reduces the number of pial-to-dural connections, thereby limiting communication between the veins intrinsic to the brain and the systemic circulation (Mack *et al.*, 2009).

2.4.3.1 Human embryos of CRL = 4 mm (45 days gestational age)

A profile reconstruction of a 4mm embryo (45 days gestational age) can be seen in figure 2.6 below. The semilunar ganglion, the otic vessels and the neural tubes are outlined on figure 2.6. One can also see the relation of these structures to the venous system of the head. This embryo is bent transversely on its longitudinal axis; thus, its left profile is seen as a convex surface and its right profile as a concave surface. The author of the article, GL Streeter did not trace the small venules and their connection with the capillary plexus. This capillary plexus is connected to the larger venous channels by minute anastomosing vessels. It also closely invests the neural tube and the adjacent sense organs and nerves (Streeter, 1915).

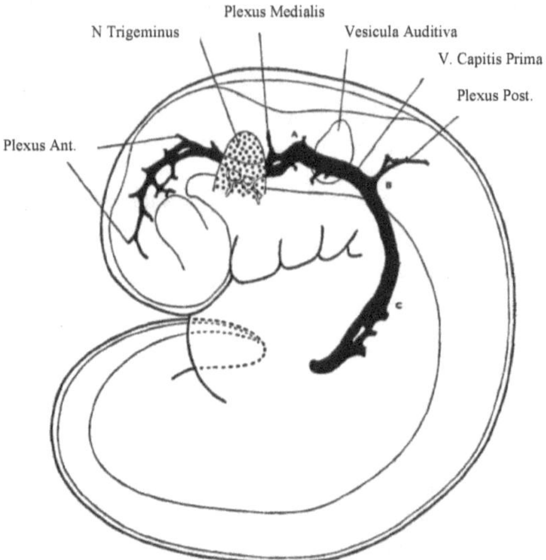

Figure 2.6: Profile of the primary head vein and its tributaries in a human embryo 4 mm long (45 days gestational age). The primary head vein is bent out of its direct course at point A by the facial and acoustic nerve-mass (sic). The segment between points B and C represents the cardinal portion of the primary head vein which will eventually become the internal jugular vein. From The development of the venous sinuses of the dura mater in the human embryo by G. Streeter, 1915, *American Journal of Anatomy,* 18, pg. 149. Used with permission.

The primary head vein constitutes the main channel and consists of a single layer of endothelial cells. The tributaries of the primary head vein unite in the region of the diencephalon. In the region of the semilunar ganglion the tributaries come together to form a main channel passing median to the ganglion. Caudally the channel passes laterally and dorsally to the acustico-facial complex. The channel continues laterally passing the otic vessel and the otic ganglion of the glossopharyngeal nerve, it then bends inward to become median and finally dorsal to the nodose ganglion of the vagus nerve. Here the channel will empty into the duct of Cuvier, which is also known as the common cardinal veins (Streeter, 1915).

The primary head vein receives mainly ventral and dorsal tributaries; of which the dorsal tributaries may be grouped as follows:

1) The anterior group from the region of the diencephalon and mesencephalon
2) The middle or cerebellar group from the region between the trigeminal nerve and the acustico-facial complex
3) The posterior or occipital group from the region of the vagus nerve rootlets

The arrangement of these dorsal tributary plexuses is significant for all the later stages.
During the later embryo stages, we find the dura and arachnoid forming; thus, allowing many of these anastomosing channels between the primary head vein and the capillaries of the brain to close off. We see that the superficial primary head vein and its tributaries become separated from the deeper vein. These deeper veins arise from and drain the capillary sheet that immediately surrounds the neural tube. However, the deeper system of veins does continue to drain into the superficial primary head vein and certain restricted places. We can now distinguish between the cerebral veins and veins of the dura mater. It is mainly the dural veins that are involved in the formation of the venous sinuses (Streeter, 1915).

We can hypothesise that if there is a persistence or non-cleavage of the primary head vein from the deeper veins, venous blood pools may be formed.

2.4.3.2 Human embryos of CRL = 14 mm long (53 days gestational age)

At 53 days gestational age the cerebral and dural veins are already separated to a considerable extent, see figure 2.7 below. Lateral to the diencephalon a large venous channel forms. This venous channel passes lateral to the otic capsules and median to the trigeminal nerve. The channel will pass through the region of the future middle ear where it will bend downwards in the neck region and finally empty into the duct of Cuvier. All the veins of the cranial region drain into this main channel. This constitutes the primary head vein. It was first described as the anterior cardinal vein. In his article Streeter, (1915) felt that the term primary head vein is adequate from the youngest stages until embryos of 20 mm (58 days gestational age) in length (Streeter, 1915).

The primary head vein has a composite origin. It is in part intrinsic to the head and in part it belongs to the anterior cardinal vein (the trunk of the primary head vein). The portion from the anterior cardinal vein will form the internal jugular vein, while the intrinsic head portion, becomes the cavernous sinus. A posterior portion may also be identified, the vena capitis lateralis, this portion will disappear completely to be replaced by another channel more dorsally (Streeter, 1915).

The tributaries of the primary head vein can be organized into three plexiform groups (figure 2.7):
1) Anterior cerebral vein - Empties into the main channel in front of the semilunar ganglion
2) Middle cerebral vein - Between the semilunar and the acustico-facial ganglia
3) Posterior cerebral vein - Caudal to the otic capsule

The posterior and middle cerebral veins each drain through a single trunk which opens into the main channel. The anterior cerebral vein, however, has multiple openings into the primary head, thus maintaining the character of the original plexus. The veins forming these three groups belong mainly to the dura mater and the tissues forming the membranous cranium. Thus, they can also be referred to as the 'anterior', 'middle' and 'posterior dural plexuses' (Streeter, 1915).

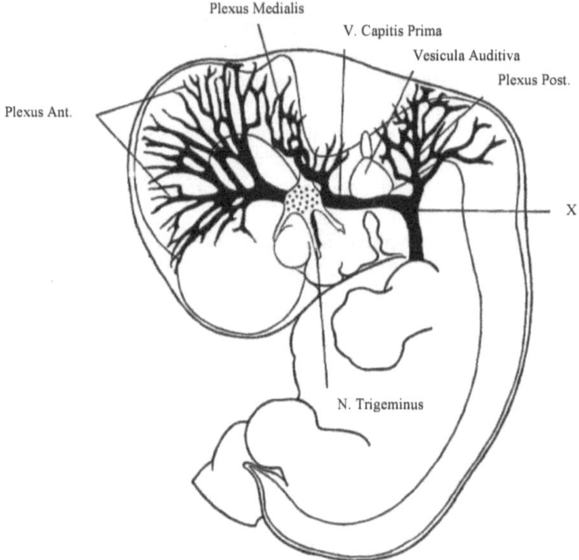

Figure 2.7: The primary head vein and its tributaries in a human embryo 13.8 mm (53 days gestational age) long. Point X marks the junction of the intrinsic head-portion of the primary head vein with the cardinal portion or internal jugular vein. From The development of the venous sinuses of the dura mater in the human embryo by G. Streeter, 1915, *American Journal of Anatomy,* 18, pg. 153. Used with permission.

Ventrally the deeper lying plexus of the neural tube is almost completely separated from the dural plexuses. The cerebral veins develop from this deeper-lying neural tube plexus. Moving more dorsally, toward the median plane, the frequency of communication between the two plexuses increases. Near the midline it becomes impossible to distinguish between the plexuses, indicating that cleavage (separation) in this region has not yet occurred (Streeter, 1915).

The larger channels seem to have favourable places where they cross over the midline. Such places include the junction of the midbrain and hindbrain, the caudal end of the roof of the fourth ventricle and lastly, along the caudal margin of the cerebral hemisphere over the diencephalon (Streeter, 1915).

We speculate that at these places where the larger channels cross the median plane, would be an ideal place to look for venous blood pools in the adult; due to the non-cleavage of the layers. These vessels are usually bilaterally asymmetrical where they cross the median line but may anastomose with the plexus of the opposite side. As a rule, the larger channels do not reach the midline other than at these three regions (Streeter, 1915).

The primary head vein also receives a large ventral tributary that lies medial to the maxillary division of the trigeminal nerve, this region also contains a number of small veins. The tissues that will in part contribute to the formation of the orbit are drained by these small veins. The future ophthalmic veins will form from these small veins (Streeter, 1915).

2.4.3.3 Human embryos of CRL = 18 mm long (56 days gestational age)

During examination of the embryo at about 14 mm in length (53 days gestational age) the primary head vein retains its course and relations; however, the patterns of the dural plexuses are in constant change. At roughly 18 mm in length (56 days gestational age) an important change occurs within the embryo. An anastomosing channel becomes established between the middle dural plexus and the posterior dural plexus. This channel passes dorsally to the otic capsule and just lateral to the endolymphatic sac. The blood from the region of the middle dural plexus drains caudally through this channel to the posterior dural plexus, as seen in figure 2.8 below (Streeter, 1915).

The trunk that originally drained the middle dural plexus into the primary head vein almost completely disappears. The blood from the posterior part of the midbrain and from the cerebellar region adopts a new channel. This channel forms dorsal to the otic capsule and drains into the posterior dural plexus. This change results in the original trunk becoming relatively small and partially breaks up into a small plexus. However, in later stages this trunk opens up again as an important channel (Streeter, 1915).

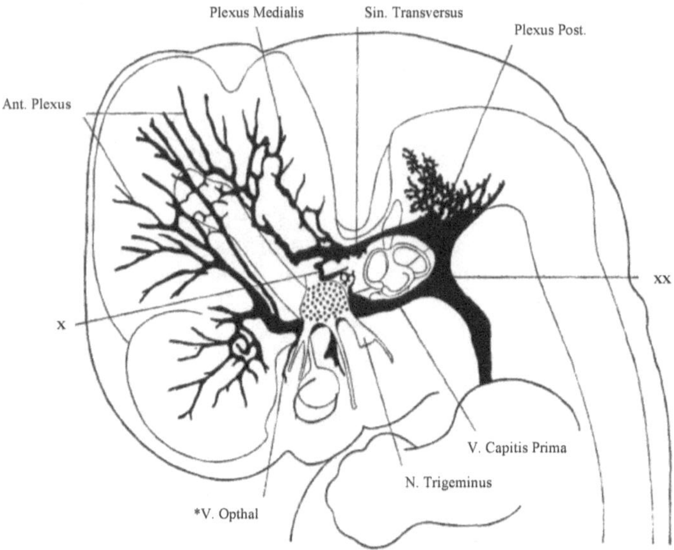

** The original author of the drawings of the embryo used in the study, G. Streeter (1915) used the abbreviation V. opthal; today the spelling is generally accepted as V. ophthalmica*

Figure 2.8: Lateral view of veins of the dura mater in a human embryo 18 mm (56 days gestational age) long. Point X marks the original trunk of the middle dural plexus; it corresponds to the superior petrosal sinus. Point XX marks the upper end of the internal jugular vein. From The development of the venous sinuses of the dura mater in the human embryo by G. Streeter, 1915, *American Journal of Anatomy,* 18, pg.156. Used with permission.

One can distinguish between the ophthalmic veins and the three dural plexuses draining into the primary head vein. The anterior dural plexus reshapes itself to form an anastomosis with the middle dural plexus. The first stage in the development of the transverse sinus is represented by the middle dural plexus draining over the otic capsule, this represents the sigmoid portion of the transverse sinus (sic). The original author Streeter (1915) used the term to distinguish between the two sides of the transverse sinus as it is needed in this specific instance to identify the orientation of the transverse sinus. There are fewer changes in the form and connections of the posterior dural plexus than in any other group of the head veins. Minor changes in its pattern do occur, other than that it just extends becoming the occipital sinus in adults. The posterior dural plexus will develop drainage channel which opens into the plexus of the tentorium cerebelli (Streeter, 1915).

We will speculate that failure of the plexus of the tentorium cerebelli to regress during development and into adulthood, could give rise to venous blood pools within the tentorium cerebelli.

The primary head vein can be subdivided into the following:
   1) The trigeminal portion that forms the cavernous sinus

2) The otic portion passing lateral to the otic capsule accompanying the seventh cranial nerve
3) The cervical portion or internal jugular vein

A new drainage channel forms dorsal to the otic capsule, this results in a decrease in volume to the otic channel. This vascular channel develops in the space dorsal to the otic channel, whereas ventro-laterally the cochlea, vestibular apparatus and the middle ear (with its contents) develops (Streeter, 1915).

2.4.3.4 Human embryos of CRL = 20/21 mm long (58/59 days gestational age)

At the 21 mm stage (59 days gestational) the veins of the head showed an intermediate arrangement between the veins in the adult and the embryonic stage. The dorsal veins resemble the embryonic stage, while the veins in the basal part of the skull resemble the adult configuration; see figure 2.9 below (Streeter, 1915).

As seen in figure 2.9, the primary head vein has transformed into its final configuration. In the region of the trigeminal nerve the primary head vein forms the cavernous sinus. Here it receives drainage from the ophthalmic veins and a large cerebral vein which drains the lateral wall of the diencephalon. This large vein mostly courses through the pia-arachnoid membranes as it forms part of the cerebral vein-system. The vein pierces the dura and continues for a short distance within the dura before it joins the cavernous sinus. It is one of the diminishing number of channels draining the cerebral venous system into the dural system. One can also see a few smaller tributaries in the region of the semilunar ganglion (Streeter, 1915).

Streeter (1915) found that in embryos smaller than 20 mm (58 days gestational age) one could identify tributaries flowing into the cavernous sinus from the cerebral hemispheres; no such tributaries were found in the larger embryos; the blood now drains caudally, toward the developing transverse sinus.

The interruption between the cavernous sinus and the internal jugular vein becomes complete; however, a small remnant may be seen as a blind channel that runs along the facial nerve. Occasionally, this channel may persist in the adult skull as the foramen of Luschka (foramen jugulare spurium), which corresponds to the exit of this channel. The vein itself, however, has never been described as persisting in humans. The vein is seen in lower forms of animals where it drains the anterior part of the brain until it empties into the internal jugular vein (Streeter, 1915).

At this stage the cavernous sinus drains upward over the semilunar ganglion and into the transverse sinus. The communication between the cavernous sinus and the transverse sinus is through a short channel resembling the original trunk of the middle dural plexus. This communication constitutes the superior petrosal sinus (Streeter, 1915).

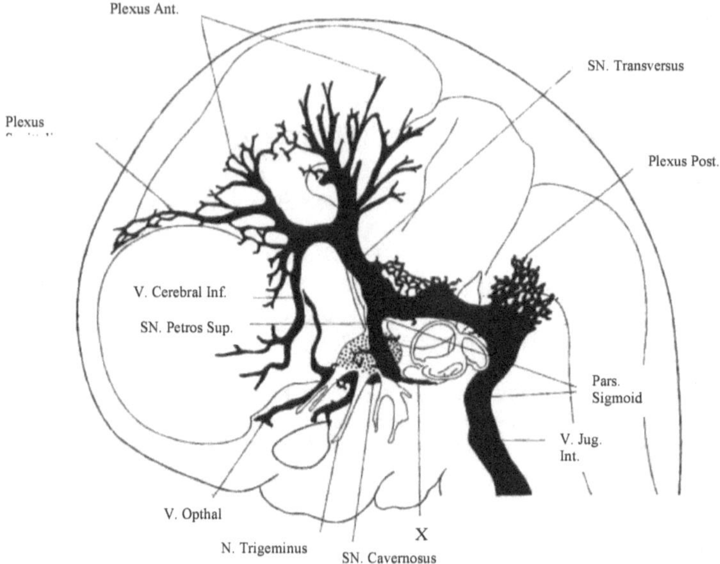

Figure 2.9: Lateral view of the veins of the dura mater in a human embryo 21 mm long (59 days gestational age). The vein marked X is the remnant of the otic portion of the primary head vein which originally connected the cavernous region with the internal jugular vein. From The development of the venous sinuses of the dura mater in the human embryo by G. Streeter, 1915, *American Journal of Anatomy,* 18, pg.159. Used with permission.

After the changes in the primary head vein a new dorsal channel drains the anterior, middle and posterior dural plexuses. This channel empties through the jugular foramen into the internal jugular vein. This channel can be seen as the transverse sinus with its sigmoid portion relations very similar to what is found in adults. However, the three dural plexuses are still of the embryonic type (Streeter, 1915).

The tentorium cerebelli consists of the dural area between the margin of the cerebellum and the cerebral hemispheres. The tentorium cerebelli forms a wedge-shaped structure which is constricted ventrally and very broad dorsally. Within its loose tissue composing the tentorium cerebelli the meshes of the dural plexuses are found (Streeter, 1915). These meshes of dural plexuses may be the origin of dural venous blood pools in the adult.

As the cerebrum and cerebellum grow it causes the region to become more compressed; thus, causing alterations in the pattern of the meshes and continued change in the venous channels contained within it. Generally, the channels radiating upwards toward the midbrain are larger than the channels nearer to the midline. The plexus also becomes finer closer to the midline forming a more intimate anastomosis with the plexus belonging to the brain wall (Streeter, 1915).

In the article by Streeter (1915), he compared the embryos of 21 mm and 18 mm in length (59- and 56-days gestational age respectively). Streeter describes a couple of characteristic changes in the pattern of the anterior dural plexus. Firstly, the anterior dural plexus drains backward into the channel dorsal to the otic capsule as it attaches itself to the middle dural plexus. Therefore, the combined anterior and middle plexuses are henceforth known only as the anterior dural plexus. Secondly, a subdivision of the anterior dural plexus differentiates between the hemispheres and along the margin of the cerebrum, this will eventually be known as the SSS (Streeter, 1915).

During the 20 mm (58 days gestational age) stage a large channel may form along the anterior margin of the anterior dural plexus. This channel formation was described by Markowski (1911) as the anterior marginal vein. The lateral surface of the cerebrum drains through a large tributary which opens into the anterior marginal vein. Markowski (1911) calls the tributary the 'lateral telencephalic vein,' of which there may be several. Bilaterally the anterior marginal veins extend forward and towards the median line. The two anterior marginal veins unit to form a plexus from which the SSS will form (Markowski, 1911). The 'anterior marginal vein' of Markowski is a part of both the sagittal plexus and the anterior dural plexus (Streeter, 1915).

According to Streeter (1915) the anterior marginal vein of Markowski can be seen as a constantly changing channel rather than a definite vein. The anterior loops of the anterior dural plexus is constantly being replaced by the more caudal loops. The larger channel which ran along the cerebral margin of the dural plexus is changing into a small mesh; with the main bloodstream forming an alternative course through a more caudal loop of the plexus (Streeter, 1915).

The migration of veins may occur in two ways:
1) Passively: The direction or position of the endothelial tube itself changes; this is due to mechanical causes arising from alteration in the environment of the vein. An example may be seen in the sigmoid portion of the transverse sinus and its change in form in the latter stages of the embryo [20 mm (58 days gestational age) in length].
2) Formation of a new channel: A vein may change its position by adopting or forming a new endothelial channel while relinquishing the original endothelial channel.

In the region of the tentorium cerebelli the embryonic plexiform character of the veins is especially favourable for migration and is repeatedly be seen in this region (Streeter, 1915).

In the embryo of 21 mm (59 days gestational age) in length, definite topographical points in the transverse sinus are noted. These are the location of the endolymphatic sac, the jugular foramen and the points of entry of the inferior cerebral veins and the superior petrosal sinus. Thus, more than half of the sinus is already established, with the terminal or jugular portion established first. The remainder of the sinus will only later assume its permanent shape later as a result of the prolonged period of growth of the cerebrum concurrently making continuous adjustments to the tentorial plexus as necessary. The proximal end of this sinus is still in its formative stage in embryo of 50 mm (82 days gestational age) long (Streeter, 1915).

2.4.3.5 Human embryos of CRL = 50 mm long (82 days gestational age)

At this stage in development the venous drainage of the cranium is established and corresponds closely to that of an adult; being separated into three systems:
- Superficial system draining the integument and soft parts

- Dural system lying between the dura and bone
- Cerebral system

Originally all three systems are outgrowths of the same capillary plexus. In embryos of 20 mm (58 days gestational age) in length the membranous and cartilaginous cranium form a separation between the dural system and the superficial veins. The superficial veins are first seen in the lower part of the head where they were originally separated from the deep veins. From here the gradually spread over the vault of the cranium forming a plexus. As seen in figure 2.10 below a few anastomoses remain between the superficial veins and the dural system. These anastomoses are the so-called emissary veins. The main drainage of the superficial system is through the external jugular vein, with the only exception being the channel maintained through the orbit (Streeter, 1915).

Upon examination of the dura in the 50 mm (82 days gestational age) long embryos, one will see that the dura largely invests close to the interior of the developing cranium and it is relatively deprived of blood vessels. This is especially seen in the frontal, temporal and lower occipital regions as well as on the base of the skull. In these regions the cartilaginous and bony cranium is more advanced in its differentiation. Except for the aforementioned, the dura projects within the cranial cavity, separated by a layer of areolar tissue from the future bony skull. The large blood channels and their tributaries can be found in this areolar tissue (Streeter, 1915).

The largest area of dura of this kind is the tentorium cerebelli; which extends from the caudal margin of the cerebral hemisphere to the cerebellum. The tentorium cerebelli can be seen narrowing as it extends toward to the base of the skull. The greatest part of the dural venous system is contained within the tentorium cerebelli (Streeter, 1915).

The cavernous sinus is formed within the meshes of the tentorium cerebelli as it widens in the region of the semilunar ganglion. A thin area of the same tissue extends posterior and medial to the otic capsule, from the cavernous sinus to the jugular region. A slender plexus of veins (the inferior petrosal sinus) extends through this area (Streeter, 1915).

The same areolar meshwork is seen along the sinuses. One should not confuse this meshwork with the developing arachnoid tissue. The areolar meshwork is separated from the arachnoid tissue by the dura. At this stage in development blood vessels supplying and draining the brain are also found in the arachnoid and subarachnoid space. These blood vessels are numerous in certain regions, for example along the ventral parts of the mid- and hindbrain and in the region of the Sylvian fissure (lateral sulcus). These cerebral vessels are distinct and separated from the dural blood channels except for a few points where they empty into the large dural channels, as in an adult (Streeter, 1915). Thus, the connection between the cerebral system and the dural system transforms from an anastomotic meshwork into isolated larger veins.

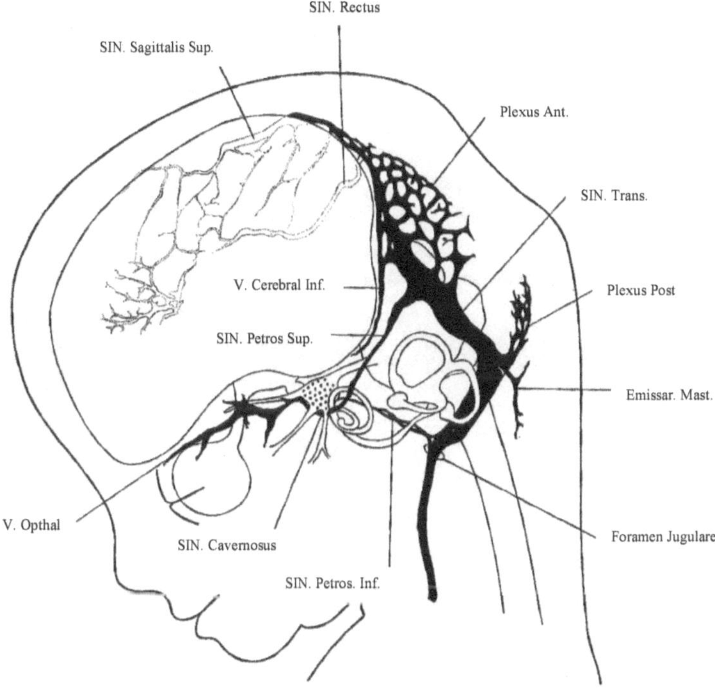

Figure 2.10: Lateral view of the dural veins in the human embryo 50 mm long (66 - 67 days post-fertilisation). The outlines of the veins of the falx cerebri can be seen through the cerebral hemisphere. From The development of the venous sinuses of the dura mater in the human embryo by G. Streeter, 1915, *American Journal of Anatomy,* 18, pg. 165. Used with permission.

By examining the figure 2.10 above, one sees that the configuration of the dural venous system follows the adult system in most respects. The cavernous sinus is not yet as developed as in the adult. The sinus is located median and ventral to the semilunar ganglion with the large ophthalmic tributaries anteriorly, communicating with the main bloodstream via the inferior and superior petrosal sinuses caudally (Streeter, 1915).

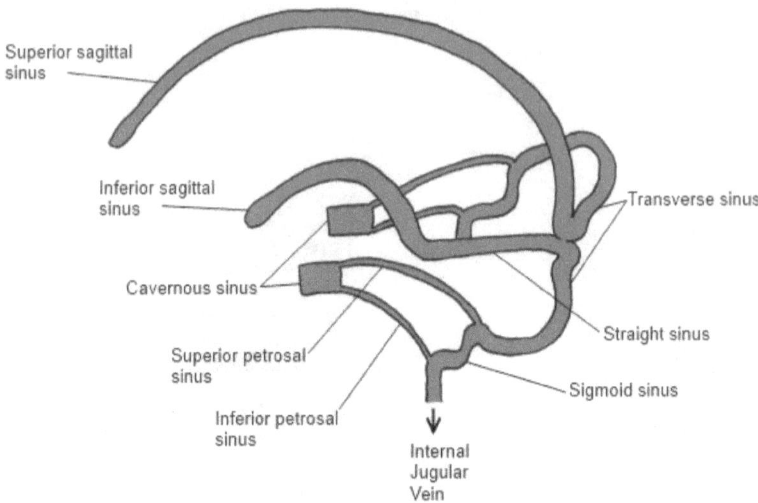

Figure 2.11: Simplified representation of the dural venous sinuses. Retrieved December, 22, 2019 from   http://www.cambridgequestions.co.uk/DisplayQuestion.aspx?id=398. Used with permission.

The terminal or jugular portion of the transverse sinus develops first. Where the superior petrosal sinus enters the jugular fossa, the transverse sinus consists of a single large channel. The transverse sinus is now seen to have the same tributaries and relations as in the adult. The proximal part of the transverse sinus is not as well established as the terminal part. Along the dorsal margin of the transverse sinus a large capillary meshwork can still be seen, this meshwork indicates that the blood channels here are still in their formative stages. These blood channels may still be referred to as the remainder of the anterior dural plexus (Streeter, 1915).

The inferior cerebral vein drains into the main channel that forms along the anterior margin of the remainder of the anterior dural plexus. In adjusting to the growth of the hemispheres this main channel is seen migrating posteriorly and assumes a more horizontal course. Thus, this change in direction, with the growth in the main channel (length and diameter) at the expense of the formative meshwork remains to be completed before the adult configuration can be considered established. Variations in the region of the confluence of sinuses in the adult may be understood as variations in the channel selection through this tentorial meshwork (Streeter, 1915). Refer to figure 2.11 for a simplified representation of the dural venous sinus in the adult.

26

The falx cerebri is composed of a flattened sheet of dura and with its vascular meshwork it is directly continuous with the tentorium cerebelli. For embryos 50 mm (82 days gestational age) in length two of its permanent channels can already be seen; these channels belong to the dural sinus system. They are the straight and the SSS (Streeter, 1915).

2.5 Summary of the embryological development of the venous system

In summarising the development of the dural venous sinuses the following groupings can be made (Streeter, 1915):

1) Establishing the primary arrangement for the drainage of the head
2) Separating the veins of the head into two and finally three separate layers (the middle layer constitutes the dural veins)
3) Changes in the dural channels due to the environmental changes in the region of the middle and internal ear
4) The dural channels adjust as the brain grows and changes
5) The histological changes in the vein walls that convert them into adult sinuses

Illustrations of the successive stages can be seen in figures 2.12 - 2.19 below.

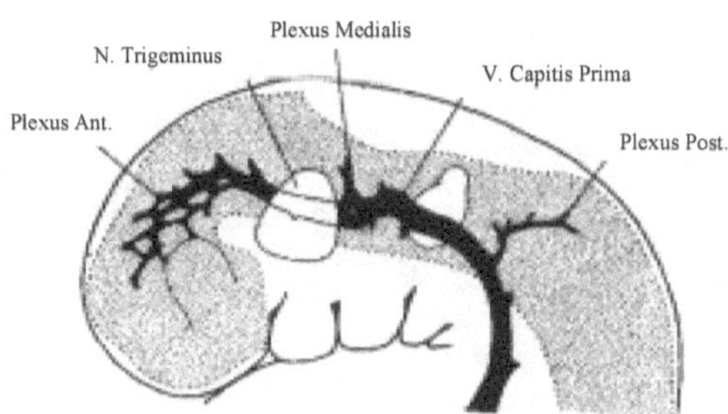

Figure 2.12: Drawing of an embryo at 4 mm CRL (45 days gestational age). From The Development of the Venous Sinuses of the Dura Mater in the Human Embryo by G.L Streeter, 1915, *American Journal of Anatomy,* 18, pg. 175. Used with permission.

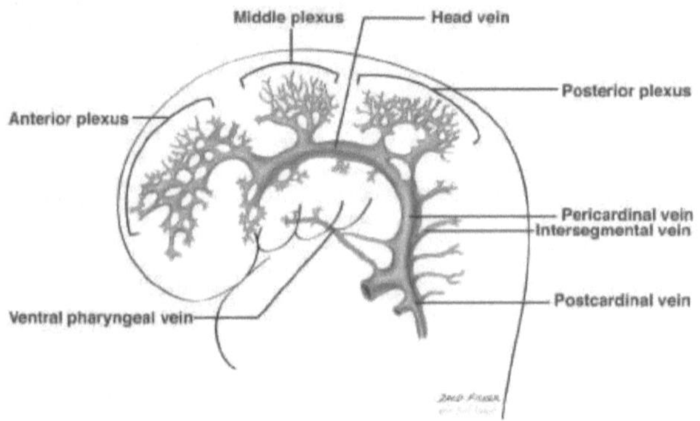

Figure 2.13: Drawing of the embryo at 4 – 6 mm CRL (45 – 47 days gestational age). From Anatomy, Imaging and Surgery of the Intracranial Dural Venous Sinuses; (pg.2), R.S. Tubbs *et al.,* 2019, St. Louis, Missouri: Elsevier. Used with permission.

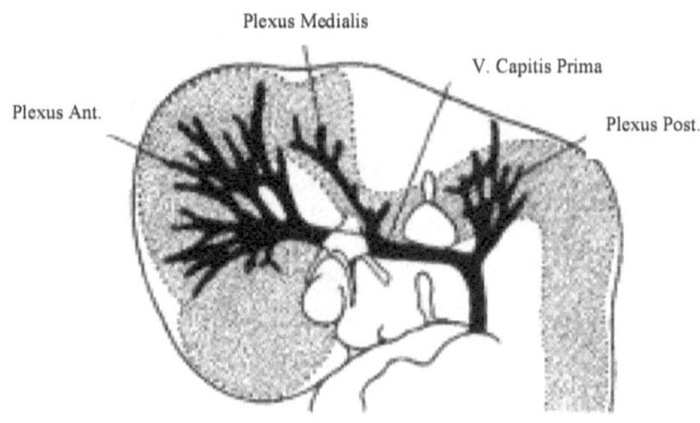

Figure 2.14: Drawing of an embryo at 14 mm CRL (53 days gestational age). From The Development of the Venous Sinuses of the Dura Mater in the Human Embryo by G.L Streeter, 1915, *American Journal of Anatomy,* pg. 175. Used with permission.

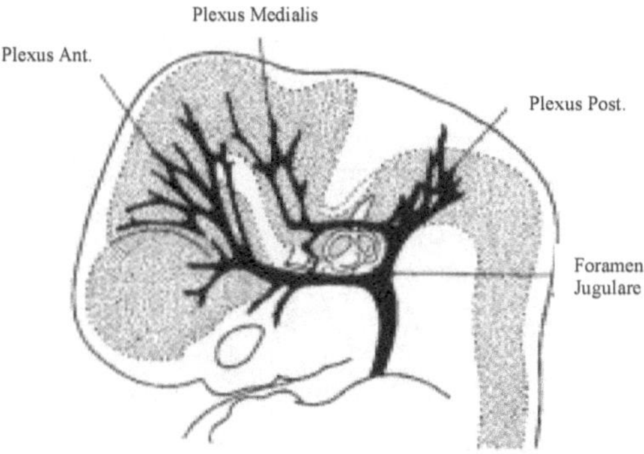

Figure 2.15: Drawing of an embryo at 18 mm CRL (56 days gestational age). From The Development of the Venous Sinuses of the Dura Mater in the Human Embryo by G.L Streeter, 1915, *American Journal of Anatomy,* pg. 175. Used with permission.

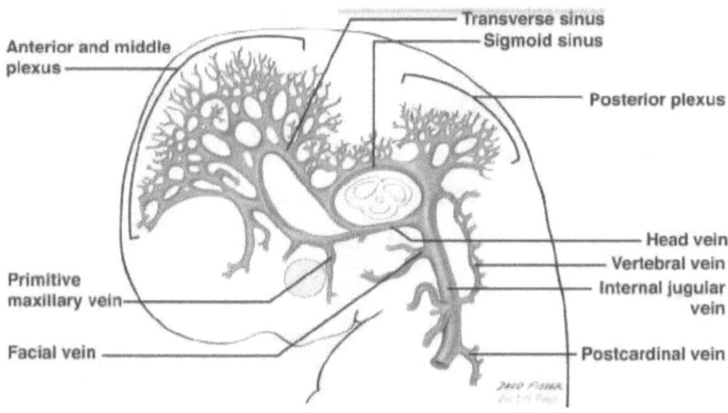

Figure 2.16: Drawing of the embryo at 16 – 18 mm CRL (55 – 56 days gestational age). From Anatomy, Imaging and Surgery of the Intracranial Dural Venous Sinuses; (pg. 3), R.S. Tubbs *et al.,* 2019, St. Louis, Missouri: Elsevier. Used with permission.

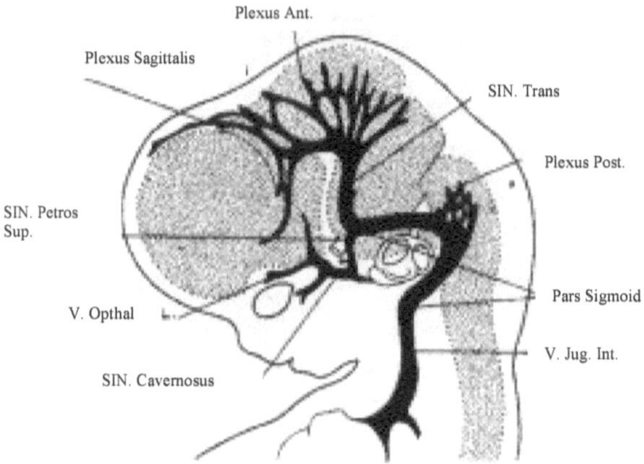

Figure 2.17: Embryo of 21 mm CRL (59 days gestational age). The Development of the Venous Sinuses of the Dura Mater in the Human Embryo by G.L Streeter, 1915, *American Journal of Anatomy,* pg. 175. Used with permission.

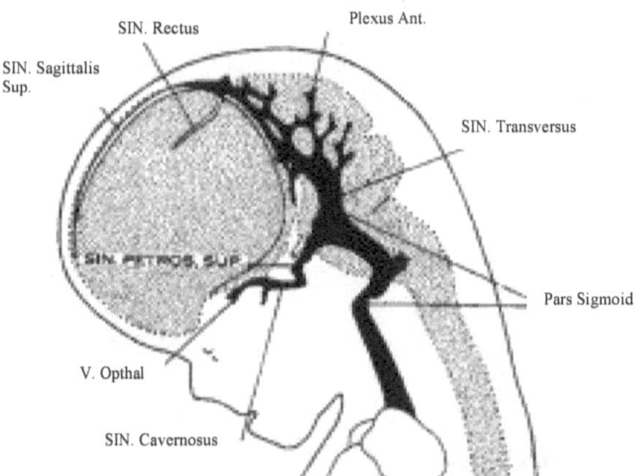

Figure 2.18: Embryo of 35 mm CRL (71 days gestational age). The Development of the Venous Sinuses of the Dura Mater in the Human Embryo by G.L Streeter, 1915, *American Journal of Anatomy,* pg. 175. Used with permission.

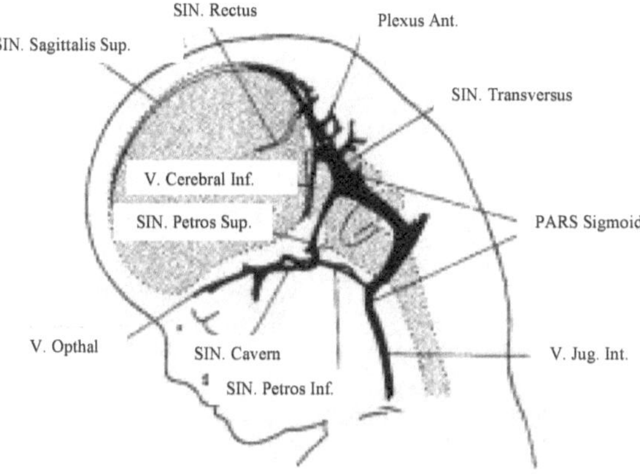

Figure 2.19: Embryo of 50 mm CRL (82 days gestational age). The Development of the Venous Sinuses of the Dura Mater in the Human Embryo by G.L Streeter, 1915, *American Journal of Anatomy,* pg. 175. Used with permission.

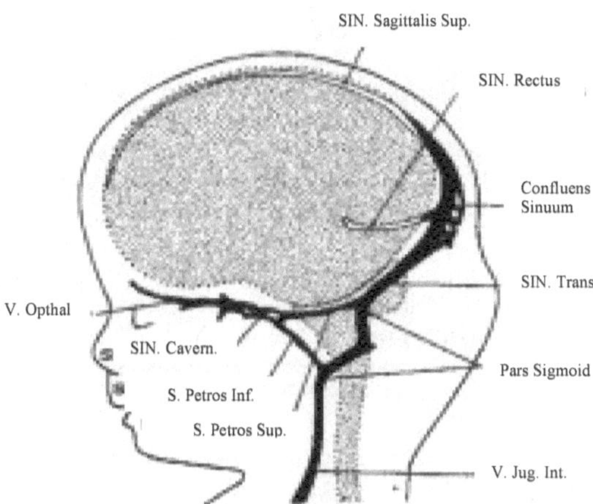

Figure 2.20: Embryo of 80 mm CRL (97 days gestational age). The Development of the Venous Sinuses of the Dura Mater in the Human Embryo by G.L Streeter, 1915, *American Journal of Anatomy*, pg. 175. Used with permission.

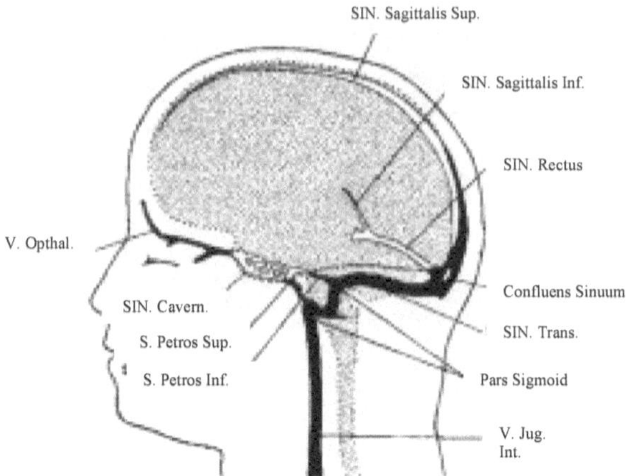

Figure 2.21: Dural venous sinus configuration in an adult. The Development of the Venous Sinuses of the Dura Mater in the Human Embryo by G.L Streeter, 1915, *American Journal of Anatomy*, pg. 175. Used with permission.

Figures 2.12 - 2.20: Simplified sketches of the dural veins illustrating the main stages of their development in the human embryo from 4mm in length to birth.

Figure 2.12 and 2.13 shows the primary arrangement of the capillaries of the head consisting of the primary head vein. The primary head vein begins in the midbrain and opens into the duct of Cuvier, on its course it runs caudally along the neural tube. The caudal portion of the primary head vein constitutes the anterior cardinal vein and will eventually become the internal jugular vein. The intrinsic vein to the vagus nerve also arises from the primary head vein. Blood from the capillary sheet investing the neural tube drains through anastomosing venous loops into a continuous channel formed by the portions of the primary head vein. Gradually these loops will arrange into the anterior, middle and posterior dural plexuses (Streeter, 1915).

Figure 2.13 and 2.14 illustrates the three head plexuses forming. The dura mater and the arachnoid space develops. The primary head vein and the three tributary plexuses are separated from the subjacent vessels arising from and draining the capillary sheet which invests the neural tube. Thus, the primary head vein and its tributaries are distinguishable from the cerebral veins as they become established as a true dural system (Streeter, 1915).

At a later stage the diploic veins form as a subdivision of the dural system. One can distinguish between three separate venous systems of the head:

1) The superficial layer - belonging to the integument and its soft parts
2) The middle layer - belonging to the dura and diploë
3) The deep layer of cerebral veins - belonging to the brain

The middle layer (dural system) contributes to the formation of the dural sinuses.

Figure 2.14 - 2.20 show the changes in the dural veins after the drainage patterns of the head have become established. These changes are largely due to the mechanical factors, they include changes in the growth and alteration of the developing brain as well as changes in the region of the cartilaginous capsule of the labyrinth (Streeter, 1915).

In embryos between 35 and 50 mm (71- and 82-days gestational age) in length (figure 2.19 and 2.20) the main channel of the anterior dural plexus can be seen. This main channel will eventually form the transverse sinus. Upon examination of the sigmoid portion, it is seen that the transverse sinus forms a relatively straight line with the internal jugular vein. Between the 50 mm (82 days gestational age) stage and adult, the transverse sinus undergoes a caudal flexion until it comes to lie at an angle of 90 degrees with the internal jugular vein (Streeter, 1915).

Figure 2.20 illustrates the transverse sinus becoming more established and the dural plexus becoming relatively smaller. The superior and straight sinuses participate in the change in position of the transverse sinus, meeting at a point, known as the confluence sinuum; which represents the last trace of the anterior dural plexus (Streeter, 1915).

Figure 2.21 illustrates the final arrange of the dural venous sinuses in the adult.

2.6 Anatomy of the venous drainage patterns

The veins are located closer to the bone than the arteries; they may even occupy their own grooves, which causes them to be particularly liable to tearing in cranial fractures, thus their anatomical relationships should be carefully examined (Standring, 2008).

2.6.1 Venous system in the child

Information regarding magnetic resonance venograms (MRV) of intracranial veins and sinuses of children are very limited. The anatomy of the normal venous system needs to be understood in order to obtain enough background information for future studies into the variations of venous structures in malformations of the brain (Widjaja & Griffiths, 2004).

The normal anatomy as described by Widjaja & Griffiths, (2004) will be discussed in relation to the variations of the venous drainage system in children under section 2.8.1.

2.6.2 Venous system in the adult

The cerebral venous system consists of two basic parts: a deep and superficial system (Uddin *et al.,*2006). These venous systems have thin walls without muscular tissue and valves. The veins pierce the arachnoid and the inner layer of the dura to drain into the dural sinuses (Griffiths, 2008). The surfaces of the cerebral hemispheres are drained by the SSS and cortical veins, together they comprise the superficial system. The deep system comprises the straight, sigmoid and transverse sinuses along with the deeper cortical veins. Both the superficial and deep systems drain primarily to the internal jugular veins (Uddin *et al.*, 2006).

2.6.2.1 The superficial system

The superficial cerebral veins can be divided into three collecting systems:
1) The mediodorsal group, draining to SSS and the SS

2) The lateroventral group draining into the lateral sinus (In the current study the term lateral sinuses is used to refer to the transverse and sigmoid sinus together)
3) The anterior group draining into the cavernous sinus

These superficial veins are connected by the great anastomotic vein of Trolard. The vein of Trolard connects the superficial middle cerebral veins to the SSS. The middle cerebral veins themselves are connected to the transverse sinus by the vein of Labbé (Uddin *et al.*, 2006).

Veins of the cerebral hemispheres:

The surfaces and interior of the cerebral hemispheres are drained by the internal and external cerebral veins. One may divide the external cerebral veins into three groups: superior, middle and inferior as described below (Griffiths, 2008).

- **Superior cerebral veins:** There are eight to twelve superior cerebral veins which drain the medial and superolateral surfaces of the hemispheres. They mainly follow the sulci; however, some may cross over the gyri. The superior cerebral veins receive small veins from the medial surface of the cerebral hemispheres before they drain into the SSS. In the anterior part of the cerebral hemispheres the superior cerebral veins join the SSS at almost right angles. The posterior veins are directed against the direction of blood flow; this configuration may assist in resisting collapse if the intracranial pressure is raised. The posterior veins are also larger than the superior cerebral veins (Griffith, 2008).

- **Middle cerebral veins:** The middle cerebral veins follow the lateral fissure and opens into the cavernous sinus. The superficial middle cerebral vein drains most of the lateral surface of the cerebral hemispheres. The superior sagittal and cavernous sinuses are connected by the vein of Trolard (superior anastomotic vein), which runs posterosuperior between the superficial middle cerebral vein and the SSS. The vein of Labbé (inferior anastomotic vein) will in turn connect the superficial middle cerebral vein to the transverse sinus. The vein of Labbé courses over the temporal lobe. The deep middle cerebral vein drains the insular region, it joins the anterior cerebral and striate veins to form the basal vein. The anterior cerebral and striate veins drain the areas which approximately correspond to the area supplied by the anterior cerebral artery. The central branches which enter the anterior perforating substance and midbrain, receive tributaries from this vicinity and join the great cerebral vein. The inferior cerebral veins join the superior cerebral veins as they drain into the SSS. The basal and middle cerebral veins anastomose with the veins from the temporal lobe and they drain to the transverse, superior petrosal and cavernous sinuses (Griffiths, 2008).

- **Inferior cerebral veins:** The basal vein begins at the anterior perforated substance where it is formed by the union of the deep cerebral vein, anterior cerebral vein and the striate veins. The basal veins receive tributaries from the inferior horn of the lateral ventricle, midbrain, para-hippocampal gyrus and the interpeduncular fossa along its course posteriorly and around the cerebral peduncle. The basal vein terminates where it drains into the great cerebral vein (Griffiths, 2008).

Veins of the posterior cranial fossa:

The veins of the posterior fossa may be divided into three groups (Uddin *et al.*, 2006):
1) The superior group draining into the Galenic system

2) The anterior group draining into the petrosal sinuses
3) The posterior group draining into the confluence of sinuses and the neighbouring transverse sinuses

Angiographic diagnosis of any occlusion of the posterior cranial veins is extremely difficult due to their variable course (Uddin *et al.*, 2006).

Veins of the cerebellum:

The veins of the cerebellum may be seen on the surface of the cerebellum where they form inferior and superior groups. These veins drain mainly into the sinuses adjacent to them or from the superior surface into the great cerebral vein (Griffiths, 2008).

The superior cerebellar veins may course anteromedially to the SS or great cerebral vein as they cross the superior vermis; or course laterally to the transverse and superior petrosal sinuses (Griffiths, 2008).

A small median vessel which runs posteriorly on the inferior vermis of the cerebellum may be seen as part of the inferior cerebellar veins. This small median vessel may open into the sigmoid or SS (Griffiths, 2008).

2.6.2.2 The deep system

Angiographically speaking the deep cerebral veins are more important than the superficial cerebral veins. The venous angle is part of the deep system and is the point of union between the following veins:
- The superior choroid vein (from the choroid plexus of the lateral ventricle)
- The septal vein (from the septum pellucidum in the anterior horn of the lateral ventricle)
- The thalamostriate vein (runs in the thalamostriate groove, between the thalamus and lentiform nucleus in the floor of the lateral ventricle)

These veins unite posterior to the interventricular foramen (of Monro) to form the internal cerebral vein. (Uddin *et al.*, 2006).

The great cerebral vein of Galen is formed by the union of the two internal cerebral veins as the run posteriorly in the roof of the third ventricle. Due to the location of these veins (within 2 mm of the midline of the brain) they are diagnostically important to identify midline shifts (Uddin *et al.*, 2006).

The vein of Galen is short and thick in appearance as it passes posterosuperior to the splenium of the corpus callosum. It drains to the anterior end of the SS (where it unites with the inferior sagittal sinus), and receives the basal and posterior fossa veins (Uddin *et al.*, 2006).

The basal vein of Rosenthal is formed by the union of the striate, middle cerebral and anterior cerebral veins. Bilaterally, the basal vein passes around the midbrain to join the great cerebral vein (Uddin *et al.*, 2006).

Thus, blood from the basal ganglia and the deep white matter of the brain are drained by the basal veins of Rosenthal and the internal cerebral veins. The basal vein of Rosenthal and the internal cerebral vein ultimately join to form the vein of Galen, which drains to the SS. Except for anatomical variations of the basal vein, the deep system is rather consistent in its structure compared to the superficial venous system. Hence, thrombosis in the deep system is easier to recognise (Uddin *et al.*, 2006).

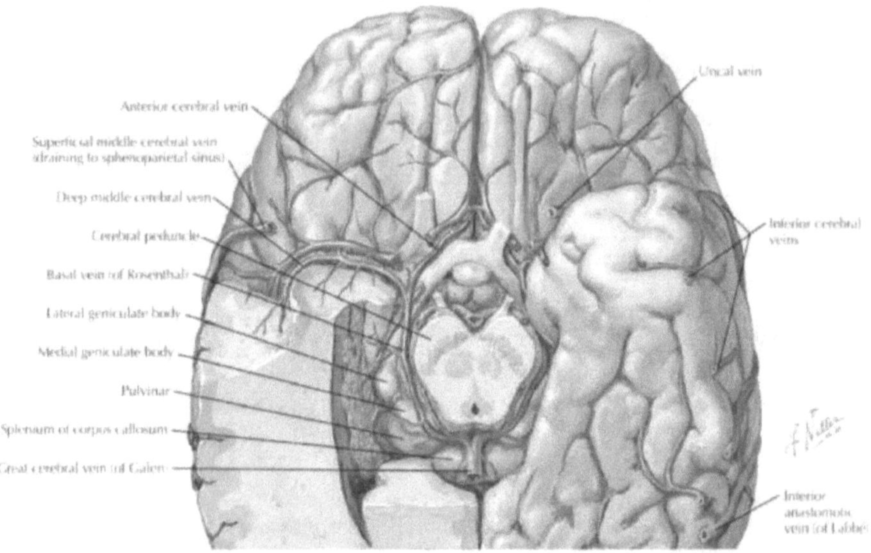

Figure 2.22: Image showing the deep venous system of the adult. Retrieved 04 January 2020, form *Netter's Atlas of Neuroscience* (pg.118), D. L. Felten, 2016, Philadelphia: Elsevier. Copyright 2016 by Elsevier. Used with permission.

Veins of the brain stem:

Deep to the arteries a venous system is formed by the veins of the brain stem. The veins from the medulla oblongata drain into the venous drainage system of the spinal cord or into the adjacent dural venous sinuses. The veins may also drain into the radicular veins which accompany the last four cranial nerves. Ultimately the veins of the medulla oblongata drain either into the occipital or inferior petrosal sinus or even the superior bulb of the internal jugular vein, refer to figure 2.22 above (Griffiths, 2008).

The anterior and posterior medullary veins run along the anterior median fissure and posterior median sulcus. The pontine veins may drain into the cerebellar veins, basal vein, transverse sinuses, petrosal sinuses or the venous plexus of the foramen ovale. The pontine veins may include a lateral and a median vein on each side. The veins from the midbrain may join the basal vein or the great cerebral vein (Griffiths, 2008).

2.6.3 Meningeal veins

Beginning from plexiform vessels in the dura mater, the meningeal veins drain into efferent vessels in the outer dural layer which connect with lacunae associated with some of the cranial sinuses. The meningeal veins include the diploic veins and the middle meningeal, refer to figure 2.24 below (Standring, 2008).

2.6.3.1 Middle meningeal veins

The frontal branch of the middle meningeal vein crosses the floor of the middle cranial fossa. It runs from either the foramen spinosum or the foramen ovale to the pterion. It is usually seen as two parallel channels accompanying the middle meningeal artery. Subsequently the vein passes cranially along the anterior margin of the parietal squama and empties into the venous lakes of the SSS. The anterior branches of the middle meningeal vessels are, for a short distance, contained within a bony canal, the sphenoparietal canal (of Trolard) as they course under the most lateral aspect of the lesser sphenoid wing. These anterior branches leave the sphenoparietal canal to enter a groove on the internal surface of the parietal squama (Standring, 2008).

Before entering the sphenoparietal canal, the anterior branch of the middle meningeal vein connects with the sinus of the lesser sphenoid wing. The sinus of the lesser sphenoid wing is connected medially to the anterior and superior aspect of the cavernous sinus with a channel that crosses over the superior ophthalmic vein to reach the cavernous sinus (Standring, 2008).

Crossing the foramen spinosum to end in the pterygoid venous plexus is the parietal (posterior) trunk of the middle meningeal vein. The frontal (anterior) trunk may also reach the plexus via the foramen ovale, or it may end in the sphenoparietal or cavernous sinus as described previously (Standring, 2008).

The middle meningeal vein (refer to figure 2.23 below) frequently contains arachnoid granulations; it also receives meningeal tributaries and small inferior cerebral veins. It connects with the diploic and superficial middle cerebral veins (Standring, 2008).

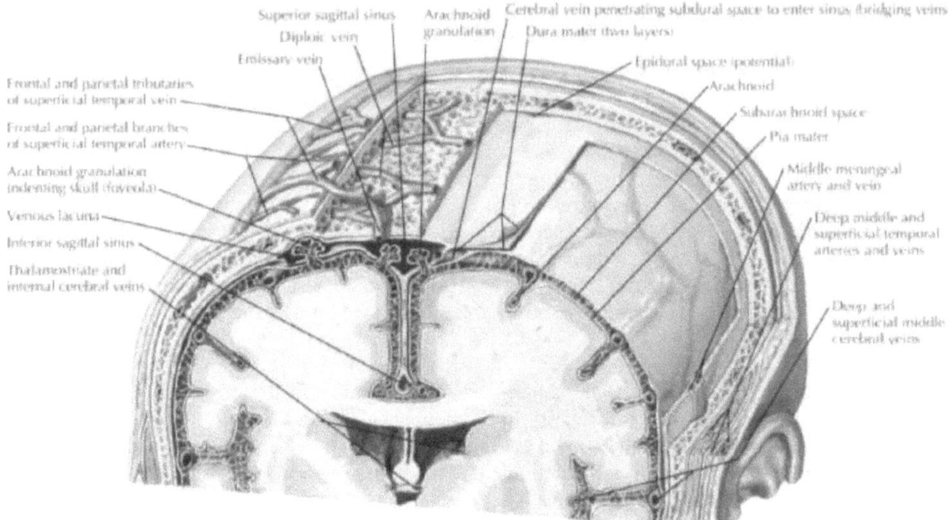

Figure 2.23: Image showing the meningeal veins of the adult. Retrieved 04 January 2020, form *Netter's Atlas of Neuroscience* (pg. 117), D. L. Felten, 2016, Philadelphia: Elsevier. Copyright 2016 by Elsevier. Used with permission.

2.6.3.2 Diploic veins

Occupying the diploë in the cranial bones are the larger, thin-walled diploic veins. Four main trunks can be described:
i.   Frontal which opens into the supraorbital vein and SSS
ii.   Anterior temporal opening into the sphenoparietal sinus and one of the deep temporal veins through an aperture in the great wing of the sphenoid; it is mainly confined to the frontal bone
iii. Posterior temporal opening into the transverse sinus through either an opening at the mastoid angle of the parietal bone or through the mastoid foramen; it is in the parietal bone
iv.   Occipital which opens externally into the occipital vein, or internally into either the transverse sinus or the confluence of sinuses; it is the largest of the four trunks and is confined to the occipital bone

2.6.4 Emissary veins

The emissary veins cross cranial apertures and form the connection between the intracranial venous sinuses and the extracranial veins. These connections are of clinical importance due to their role in determining the spread of infections from extracranial foci to venous sinuses (Standring, 2008).

The following emissary veins may be observed (refer to figure 2.24 below):

i. The mastoid emissary vein which connects the sigmoid sinus with the posterior auricular or occipital veins

ii. The parietal emissary vein connects the SSS with the veins of the scalp; it may be found in the parietal foramen

iii. The venous plexus of the hypoglossal canal may occasionally be a single vein; it connects the sigmoid sinus and the internal jugular vein

iv. The posterior (condylar) emissary vein which connects the sigmoid sinus and veins in the suboccipital triangle via the posterior (condylar) canal

v. The occipital emissary vein connects the confluence of sinuses with the occipital vein through the external occipital protuberance; it may also receive the occipital diploic vein

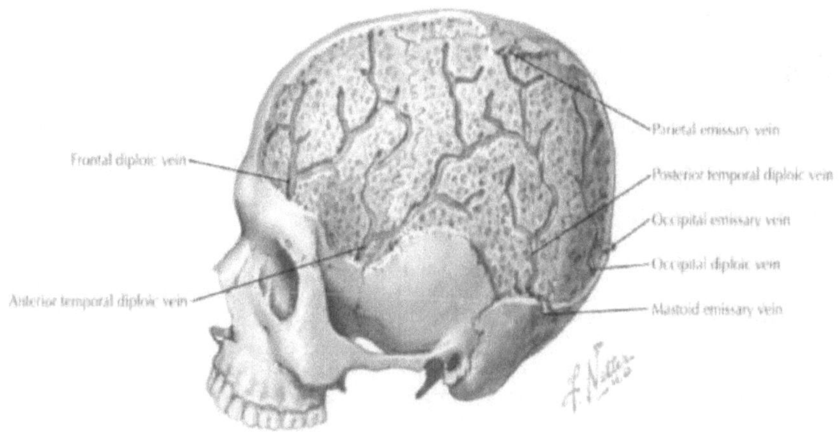

Figure 2.24: Image showing the diploic and emissary veins of the adult. Retrieved 04 January 2020, form *Netter's Atlas of Neuroscience* (pg. 117), D. L. Felten, 2016, Philadelphia: Elsevier. Copyright 2016 by Elsevier. Used with permission.

The venous plexus of the foramen ovale (a plexus of emissary veins) connects the cavernous sinus to the pterygoid plexus via the foramen ovale. Two or three small veins can be seen traversing the foramen lacerum connecting the cavernous sinus with the pharyngeal veins and pterygoid plexus. The cavernous sinus, pharyngeal veins and pterygoid venous plexus are also connected by a small vein in the emissary sphenoidal foramen (of Vesalius) also known as the foramen venosum (Standring, 2008).

The cavernous sinus and the internal jugular vein are connected by the internal carotid venous plexus, which passes through the carotid canal. The transverse sinus is connected to the external jugular vein by the petrosquamous sinus. The foramen caecum (which is patent in about 1% of adults) may be traversed by a vein which connects the nasal veins with the SSS (Standring, 2008).

Variably developed veins around the foramen magnum (the marginal sinuses) connect with the occipital sinus. Thus, the occipital sinus also connects with the vertebral venous plexuses; this provides an alternative pathway for venous drainage in case the internal jugular vein is blocked or ligated. One may also potentially consider the ophthalmic veins as emissary veins; as they connect intracranial to extracranial veins (Standring, 2008).

2.6.5 The Falcine plexus

The current medical literature includes few statements regarding veins that may be present within the falx cerebri. In the study by Tubbs *et al.,* (2007) the authors found that these veins range in size from 0.5 mm to 1.1 cm, all of which communicated with the inferior sagittal sinus. These veins are known as the falcine veins and collectively form the falcine venous plexus. The falcine veins may be grouped into three distinct types:

- **Type I:**   Veins which had no communication with the SSS
- **Type II:**   Veins had limited communication with the SSS
- **Type III:**   Veins had a significant communication with the SSS

Tubbs *et al.,* (2007) found that most of the falcine veins are located in the posterior one-third of the falx cerebri, and specifically in the inferior two-thirds of this region.

2.7 Dural venous sinuses

It is important to understand the normal anatomy of the venous structure of the brain, its variations and development before attempting to study the abnormal structures and malformations of the brain (Widjaja & Griffiths, 2004).

When discussing the venous system of the brain, it is imperative to include the anatomy of the cerebral venous sinuses, as variable as they might be (Uddin *et al.,* 2006).

Traditionally the dura has been viewed as a supportive structure containing the dural sinuses and covering the brain, but otherwise devoid of vessels. Contained within the dura is an inner vascular plexus which is smaller in the adult than in the infant. This inner plexus likely plays a role in CSF absorption, particularly in the infant whose arachnoid granulations are not completely developed (Mack *et al.,* 2009).

2.7.1 The superior sagittal sinus

During the 21 mm stage (59 days gestational age) of the embryo the SSS starts to form as a subdivision of the anterior plexus. The sagittal plexus is noted to have two characteristic features (see figure 2.25 below):

1) There is an enlargement in certain portions of the sagittal plexus, although it is a continuous channel. It forms a series of small lakes connected by narrow channels (a single, definite SSS has not yet been established);
2) The sagittal plexus is distinctly asymmetrical and tends to drain more freely to the one side.

Figures 2.25 and 2.26 below show a more definite and simpler channel system; one might refer to this as the start of the SSS in the embryo. However, as the channels are still in the form of a plexus the term, 'plexus sagittalis' is retained (Streeter, 1915).

According to the paper published by HM Evans in 1909, the primary capillary plexus 'creeps' up each side of the forebrain, as seen in the forebrain of an 8 mm pig embryo. He labelled a portion of the dorsal margin of this plexus the primitive SSS. According to Evans (1909) it is initially paired and bilaterally symmetrical. However, according to Streeter (1915), it is not until later that one can refer to a SSS. Streeter states that it is not until the plexuses as described by Evan in 1909 fuses in the midline and forms a longitudinal network, in the meshes of which an asymmetric channel is formed that one speaks of the superior sagittal sinus.

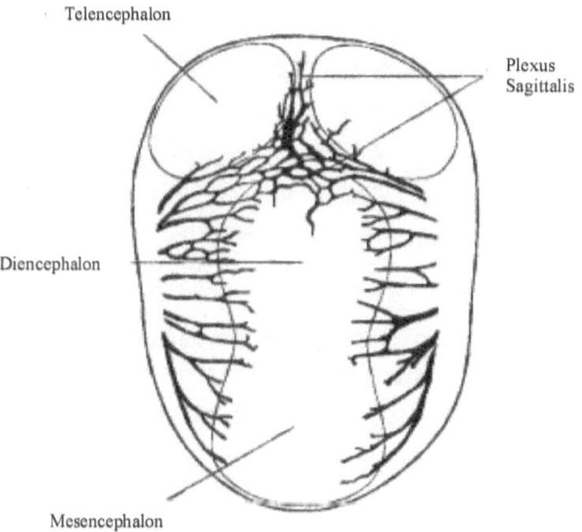

Figure 2.25: Dorsal view of a human embryo at 13.8 mm (53 days gestational age) long; showing stage 1 of the development of the SSS. The Development of the Venous Sinuses of the Dura Mater in the Human Embryo by G.L Streeter, 1915, *American Journal of Anatomy,* pg. 168. Used with permission.

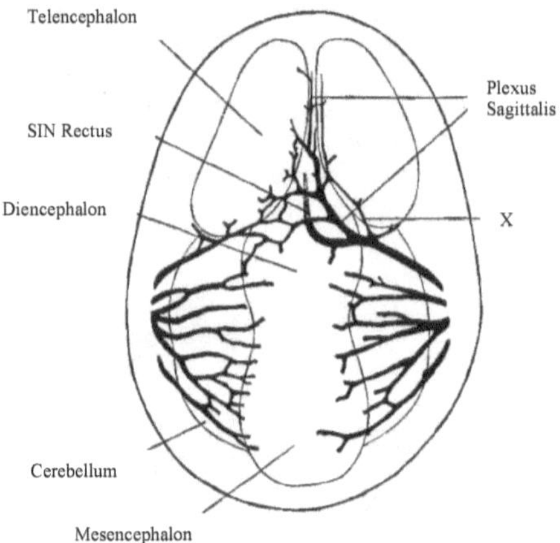

Figure 2.26: Dorsal view of a human embryo at 20 mm (58 days gestational age) long; showing stage 2 of the development of the SSS. The Development of the Venous Sinuses of the Dura Mater in the Human Embryo by G.L Streeter, 1915, *American Journal of Anatomy,* pg. 168. Used with permission.

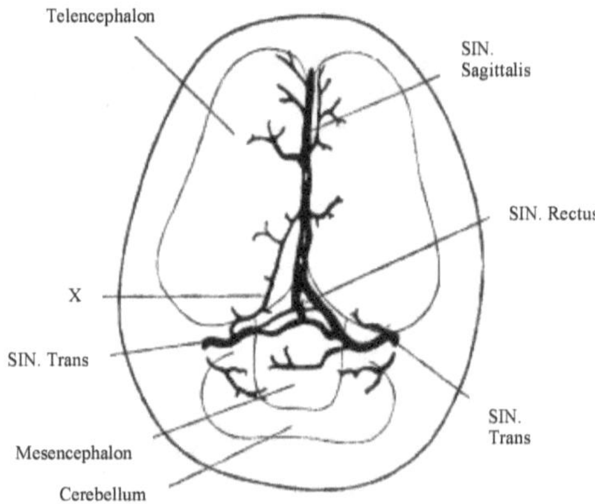

Figure 2.27: Dorsal view of a human embryo at 54 mm (82 days gestational age) long; showing stage 3 of the development of the SSS. The Development of the Venous Sinuses of the Dura Mater in the Human Embryo by G.L Streeter, 1915, *American Journal of Anatomy,* pg. 168. Used with permission.

Figures 2.25 - 2.27 show the three stages in the formation of the sagittal plexus; indicating its asymmetrical character and its conversion into the SSS. From below it receives the drainage channel from the choroid plexus which become the SS. The channels marked X are undergoing retrogression, thus they are being replaced by more caudal channels (Streeter, 1915).

Due to the growing of the cerebral hemisphere in embryos of 20 mm (58 days gestational age) in length, the dural tissue caught between the hemispheres begins to form the falx cerebri. The meshes of the sagittal plexus are found within this loose dural tissue. At this time point, one can see a larger channel opening along the dorsal midline which gives rise to the superior sagittal sinus. Connected to it with anastomosing loops is a more ventrally situated large channel that constitutes the SS. The sinus rectus extends anteriorly and drains the lower part of the falx cerebri. The choroid plexuses of the cerebral hemispheres are drained by the two converging limbs found in front of the sinus rectus (Streeter, 1915).

In embryos of 50 mm long (82 days gestational age) the SSS is well established. Cephalically is a large characteristic channel, which only needs the dural connective tissue to complete the transformation into the adult type. Caudally it still exhibits a plexiform characteristic that indicates its transitional state. According to Streeter (1915), the formation of a single channel was the result of more than one process; in some segments there is a selection of a favourable loop of the plexus which then enlarges and becomes the main channel. In other segments there is an enlargement of two or more collateral loops which subsequently fuse into one channel (Streeter, 1915).

The SSS continues to grow caudally in response to the growth of the hemispheres. As illustrated in figures 2.25 - 2.27 one can see the caudal development of the SSS is accomplished by the expense of the meshes of the anterior dural plexus, the transverse and SS also take part in this process. This caudal movement is accomplished by what was previously described as "spontaneous migration", thus the channel repeatedly shifts into a more caudal loop of the plexus, with the new loop enlarging and the old loop dwindling. Referring to figures, 2.27 and 2.28, the point marked X may be interpreted as discarded channels. The caudal development reaches its completion at the torcular Herophili (confluence of sinuses); the confluence may retain a trace of its plexiform character that was present throughout its embryonic stages (Streeter, 1915).

The asymmetry of the SSS expresses itself in the embryo as well as in the adult; this is seen in its tendency to drain more to one side of the head than to the other. This is already established in the embryo by the time it is 20 mm (58 days gestational age) in length; i.e. predominantly to the right side (Streeter, 1915). In the study done by Streeter (1915) it was noted that of 18 specimens, all but two drained to the right side. The author did not speculate as to why this asymmetry occurs. Starting at the foramen caecum in the frontal bone, the SSS runs superiorly and posteriorly toward the internal occipital protuberance. It is joined by the SS and the lateral sinus to form the confluence of sinuses (Uddin *et al.*, 2006).

The anterior part of the SSS is narrow and sometimes absent, it could be replaced by two superior cerebral veins that join posterior to the coronal suture. This variation should be kept in mind when evaluating for cerebral venous thrombosis (CVT). Most of the cerebral hemispheres are drained by the SSS (Uddin *et al.*, 2006).

2.7.2 The cavernous sinus

The cavernous sinus drains blood from the inferior parts of the frontal and parietal lobes, from the superior and inferior petrosal sinuses and the orbits are drained via the cavernous sinus. Blood from the cavernous sinus ultimately drains to the internal jugular veins (Uddin *et al.*, 2006).

The intercavernous sinuses (anterior and posterior) are dural venous sinuses which, along with the basilar plexus, connects the left and right cavernous sinuses. They lie within the anterior and posterior border of the diaphragma sellae respectively. Small venous sinuses located in the base of the pituitary fossa drain into the intercavernous sinuses; they are a cause of bleeding during transsphenoidal hypophysectomy (Hacking & Gaillard, 2018).

If both the intercavernous sinuses are present, together with the left and right cavernous sinuses they form a complete circular venous sinus (Petrosquamous sinus, retrieved from the AnatomyNext website, on 26 December 2019).

2.7.3 The straight sinus

The SS is formed by the great vein of Galen (great cerebral vein) and the inferior sagittal sinus (Uddin *et al.*, 2006).

From its origin at the falcotentorial apex, the SS runs posteroinferiorly, to where it terminates by joining the SSS and transverse sinuses to form the confluence of sinuses. It receives numerous small tributaries from the falx cerebri, tentorium cerebelli and the adjacent brain along its course (Osborn, 2013).

2.7.4 The inferior sagittal sinus

The inferior sagittal sinus is much smaller and more inconsistent when compared to the SSS (Osborn, 2013). The inferior sagittal sinus runs within the free edge of the falx cerebri, it unites with the vein of Galen to form the SS (Uddin *et al.*, 2006).

The inferior sagittal sinus collects small tributaries as it curves posteriorly along the free margin of the falx cerebri. Terminating at the falcotentorial junction, it joins with the vein of Galen (great cerebral vein) to form the SS (Osborn, 2013).

2.7.5 The lateral sinuses

The lateral sinuses extend from the confluence to the jugular bulbs and consists of a transverse and sigmoid portion. The lateral sinuses drain blood from the cerebellum, brain stem and posterior parts of the cerebral hemispheres. They are also joined by small veins from the middle ear and the diploic veins. The anatomical configuration of the lateral sinuses is numerous, which may be misinterpreted as sinus occlusion on imaging (Uddin *et al.*, 2006).

2.7.5.1 Transverse sinuses

The transverse sinuses are contained within the attachments of the tentorium cerebelli to the inner table of the skull. The sinus curves laterally from the confluence to the posterior border of the petrous part of the temporal bone, where they turn and become the sigmoid sinuses (Osborn, 2013).

Within the transverse sinuses anatomical variations are common; they are typically asymmetric, with the right side being larger than the left. One commonly finds hypoplastic or even atretic segments, one also sees filling defects caused by the arachnoid granulations and fibrous septa (Osborn, 2013).

2.7.5.2 Sigmoid sinuses

The transverse sinuses basically continue inferiorly as the sigmoid sinuses. They have gentle S-curve and descends behind the petrous part of the temporal bone, terminating by becoming the internal jugular veins. Asymmetry between the two sigmoid sinuses are commonly seen (Osborn, 2013).

Focal venous dilations at the skull base called the jugular bulbs are found between the sigmoid sinuses and the extracranial internal jugular veins (Osborn, 2013).

2.7.6 The petrosal sinuses

One can identify two petrosal sinuses; superior and inferior. The superior petrosal sinus passes over the cochlear part of the otic capsule as a long slender channel and empties into the transverse sinus. The inferior petrosal sinus on the other hand consists of a plexus of veins which pass median to the otic capsule and empties at the origin of the internal jugular vein (Streeter, 1915; Osborn, 2013; Padget, 1956).

2.7.7 Confluence of sinuses

The torcular Herophili, more commonly referred to as the confluence of sinuses is located posteriorly in the skull where the SSS terminates by joining the left and right transverse sinuses (Osborn, 2013, refer to the list of definitions on page v).

The anatomical variations of the venous sinuses draining to and from the confluence of sinuses can be grouped into three types. As described by Bisaria (1985) type I refers to individuals where the sagittal sinus drains into one transverse sinus and the SS into the other transverse sinus, with no connection between the two. Type II are individuals in which the SSS and the SS fork, and the forks from both sinuses join to form the transverse sinuses. The last type includes individuals in which the confluence of sinuses exists, which can vary from a common pool to merely a potential confluence, depending on the presence of "pads" or incomplete and complete partitions of dura mater (Bisaria, 1985).

In the study done by Bisaria (1985), the following variation were rarely seen:
- double SS draining into one transverse sinus
- the SSS dividing into three channels with two transverse sinuses on one side
- a transverse sinus originating from a tentorial vein
- a tentorial vein draining into the confluence of sinuses

Bisaria (1985) described the variations in the region of the confluence of sinuses simply by observations with the naked eye. The author studied the size, communications and anomalies of the superior sagittal, transverse and SS at the confluence of sinuses by gross dissection.
In the study done by Bisaria (1985) the author evaluated 110 adult cadavers. Of these 20 cadavers showed the SSS terminating in the right transverse sinus and the SS terminating in the left transverse sinus. In one of these 20 there were double SS running one superior to the other; these double sinuses united near the internal occipital protuberance. The SS thus formed was continuous with the left transverse sinus (Bisaria, 1985).

In seven of the cadavers the SSS was continuous with the left transverse sinus and the SS drains into the right transverse sinus. Two other cadavers had two SS; the superior SS opened into the left transverse sinus and the inferior SS into the right transverse sinus (Bisaria, 1985).

For 18 of the cadavers the SSS directly joined the confluence of sinuses. In eight of the cranial cavities "pads" of dura mater were noted in the posterior wall. Ten other specimens were seen with the SSS forked and joined with the forked SS, which then became continuous with the transverse sinus. The SSS showed partition of the dura in its terminal portion, which divided this part of the sinus into unequal channels in 34 of the specimens. A dural partition extended across the torcular Herophili to the left wall of the SS in seven of the cranial cavities. In these cases, the right channel of the SSS was continuous with the right transverse sinus and the left channel was continuous with the left transverse sinus (Bisaria, 1985).

In only four specimens the dural partition extended from the terminal portion of the SSS to the anterior wall of the right transverse sinus. Thus, causing the narrow left channel to be continuous with the left transverse sinus and the SS. The right and broader channel was continuous with the right transverse sinus. In one of the cases, the superior wall of the right transverse sinus showed a circular defect near the confluence of sinuses. In only one specimen the SSS divided into three channels. The left being continuous with a thin and superficial transverse sinus and also with the more deeply placed transverse sinus which could be traced to the sigmoid sinus. In another specimen the left transverse sinus began as a tentorial vein; with the SSS being continuous with the right transverse sinus and SS. In six specimens the tentorial veins terminated directly into the torcular Herophili (Bisaria, 1985).

2.7.8 Occipital sinus

The occipital sinus is the smallest of the dural sinuses. It is found within the attached margin of the falx cerebelli and may occasionally be paired. The occipital sinus starts at the foramen magnum as several small channels, one of which joins the sigmoid sinus, the others connecting with the vertebral plexuses. The occipital sinus drains in the confluence of sinuses (Standring, 2008).

2.7.9 Marginal sinus

The foramen magnum is encircled by the marginal sinus. Anteriorly it communicates with the basilar plexus and posteriorly with the occipital sinus. Typically, it drains to the sigmoid sinus or jugular bulb via small sinuses and may also connect extracranially to the internal vertebral venous plexus, the paravertebral or the deep cervical veins in the suboccipital region (Standring, 2008).

2.7.10 Basilar sinus and plexus

Consisting of interconnecting channels between layers of dura mater on the clivus, the basilar sinus and plexus interconnects the inferior petrosal sinuses and joins the internal vertebral venous plexus. Anteriorly it also connects with the cavernous and superior petrosal sinuses (Padget, 1956; Standring, 2008).

In the current book no distinction is made between the basilar plexus and sinus in the book. Both are used to refer to the interconnecting channels connecting the inferior petrosal sinuses.

2.7.11 Sphenoparietal sinus

The sphenoparietal sinus courses around the lesser sphenoid wing at the rim of the middle cranial fossa (Osborn, 2013). The sinus receives superficial veins from the anterior temporal lobe and receives tributaries from the superficial middle cerebral vein, middle meningeal vein and the anterior temporal diploic vein. The sphenoparietal sinus drains into the cavernous sinus (Osborn, 2013; St-Amant & Gaillard, 2018).

For an illustrated overview of the venous sinuses in the adult refer to figure 2.28 below.

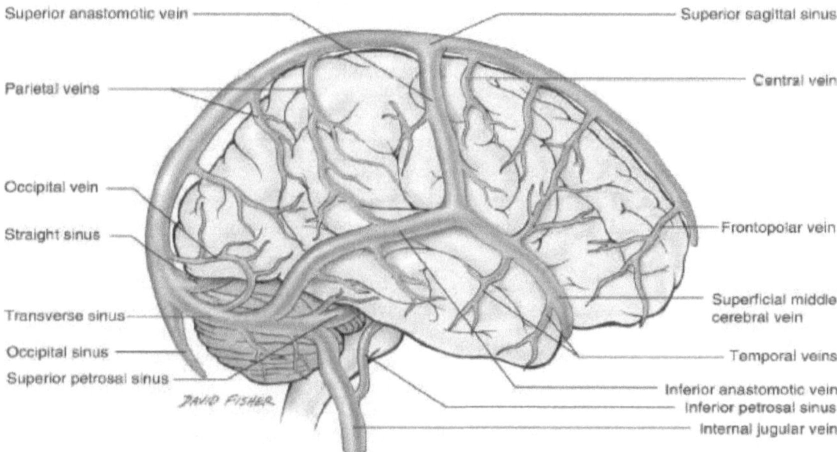

Figure 2.28: Cerebral veins and dural venous sinuses. From *Anatomy, Imaging and Surgery of the Intracranial Dural Venous Sinuses;* (pg. 11), R.S. Tubbs *et al.,* 2019, Philadelphia, Missouri: Elsevier. Used with permission.

2.8 Dominance and variation in the cerebral venous system

The understanding of the structure of the dural sinuses and their variations has increased greatly since the advances of non-invasive imaging techniques. The confluence of sinuses shows the most variations of the dural sinuses (Bayaroğullari *et al.,* 2018).

The term dominance is used to refer to the side of the brain on which the transverse sinus receives the greater volume of venous blood from the SSS.

For a summary of the definitions of dominance found in the literature, see table 2.1 below.

Table 2. 1 A summary of the definition of dominance found in the literature.

Author	Year	Age range of sample	Definition of dominance
Bayaroğullari *et al.*	2018	1 - 81 years	Classification based on the structures draining into the confluence of sinuses. No definition of dominance as authors did not look at dominance, but rather drainage patterns.
Kitamura *et al.*	2017	22 - 89 years	Measurement ratio used: right/left to determine dominance. A venous sinus was classified as dominant when the measurement ratio was more than 1.5 (dominant right) or less than 0.67 (dominant left). A venous sinus was classified as symmetrical when the measurement ratio was equal or between 1.5 and 0.67.
Goyal *et al.*	2016	19 - 86 years	The transverse sinuses were measured 1 cm from the torcula Herophili. The linear measurement was compared with the SSS. If the linear measurement was less than half the size of the SSS, it was considered hypoplastic/non dominant and if not visualised it was considered aplastic or atretic sinus.
Sharma & Sharma	2012	Not specified	Not defined
Pallewatte *et al.*	2016	Not specified	The dominant transverse sinus is the sinus which receives the SSS
Friedmann *et al.*	2011	Infancy to Adult, does not specify exact age	The dominant side for a given venous structure was defined as the one with the larger diameter of the vessel; thus, right dominance was present when the diameter on the right was larger than that on the left.
Manara *et al.*	2010	20 days to 85 years	Transverse sinus was considered asymmetrical if the diameter side-to-side difference was >10%.
Park *et al.*	2008	Not specified	Dominance was determined by calculating the ratio of the width of the right side to the width of the left side. Ratios of > or equal to 1.5 were classified as dominant on the right side, and cases with ratios <0.67 were classified as dominant on the left side. Symmetrical cases showed ratios between 0.67 and 1.49.
Surendrababu *et al.*	2006	2 - 68 years	The transverse sinuses were measured 1 cm from the torcula Heterophili. These were compared with the SSS. If the linear measurement was less than half the size of the SSS, it was considered hypoplastic/non dominant and if not visualised it was considered aplastic or absence of sinus.

Author	Year	Age range of sample	Definition of dominance
Rollins *et al.*	2005	Neonates to 17 years	A transverse sinus was considered dominant when it was larger than the contralateral transverse sinus on visual inspection of the coronal and axial MIP images. A transverse sinus was considered absent when it was absent on both the MIP and source images from the torcula to the sigmoid sinus.
Widjaja & Griffiths	2004	3 months to 17 years	A transverse sinus was said to be hypoplastic when the linear dimension of the transverse sinus was less than half the diameter of the SSS.
Ayanzen *et al.*	2000	9 days to 83 years	Not defined
Durgan *et al.*	1993	Not specified	Equal dominance defined as venous drainage equal to right and left transverse sinuses. The transverse sinus receiving the most venous drainage is said to be the dominant sinus.
Browning	1953	Adults (range not specified)	Minor dominance: The ratio between the capacities of the transverse sinuses was less than three to one. Major dominance: The ration between the capacities of the transverse sinuses were more than three to one.
Gibbs *et al.*	1934	Not specified	Not defined
Edwards	1931	Adults (range not specified)	Not defined

2.8.1 Cerebral venous system of the infant and child

Transverse and sigmoid sinuses

From around the fourth fetal month the transverse sinuses enlarge from their lateral border on each side. This enlargement progresses medially until it reaches the confluence of sinuses. With the increase in volume of the telencephalon the area where the SS joins the SSS gradually descends, resulting in the inferior inclination of the lateral portion of the transverse sinuses being less marked as the fetus develops (Okudera *et al.*, 1994).

The transverse sinuses will continue to enlarge until the sixth fetal month, when they regain a relatively even calibre. There is a transition period from the plexiform network stage to that of rapid increase and then a decrease in calibre; during which the dural sinuses in the region of the torcular Herophili and the lateral sinuses may become irregular. One might even find that the medial and lateral portions of the transverse sinus are absent. Marked hypoplasia or aplasia of major portions of the transverse sinuses may be seen. One may therefore frequently encounter unequal heights, asymmetrical sizes, irregularities or absence of the medial portion of the transverse sinuses. The lateral portion of the transverse or sigmoid sinus might be hypoplastic or absent, however hypoplasia and aplasia of the sinus is less common (Okudera *et al.*, 1994).

Widjaja & Griffiths, (2004) found unilateral absence of the transverse sinus as well as flow gaps of the non-dominant transverse sinus in three cases; which they ascribed to changes during development of the sinus.

In the study by Rollins *et al.,* (2005) they evaluated three groups of patients.
- **Group one:** Consisted of 38 patients (neonates to 24 months of age); of these 32 had normal MR images on evaluation (without pathology).
- **Group two:** Consisted of 17 patients (25 months to 5 years of age), of these 16 had normal MR on evaluation (without pathology).
- **Group three:** Consisted of 53 patients (6 to 17 years of age), 52 had normal MR images on evaluation (without pathology).

Group one had 14 (37%) patients with a dominant right transverse sinus, eight (21%) with a dominant left and 16 (42%) had codominant transverse sinuses. In group two, six (35%) patients had right transverse sinus dominance and five (30%) left sided dominance, codominant transverse sinuses were seen in 6 (35%) patients. Absent transverse sinus was seen in three (18%) patients. An absent transverse sinus was defined by an absence of signal intensity in the transverse sinus on the source images. For group three, the right transverse sinus was dominant in 26 (50%) of patients, left dominant in nine (16%) and codominant in 18 (34%) of patients. An absent transverse sinus was noted in seven (13%); however, signal intensity loss was seen in about 50% of patients with non-dominant transverse sinuses. One patient had an occipital sinus associated with the absence of a transverse sinus (Rollins *et al.,* 2005).

Rollins *et al.,* (2005) conducted a chi-square contingency analysis which showed no significant relationship between transverse sinus dominance and age when the three groups were compared (p=0.58). Some relationship was shown between transverse sinus dominance and age when group one and two were combined (patients younger than six years) and compared with group three (patients older than six years); (p=0.032). A chi-square trend analysis showed a mildly positive correlation between the absence of a transverse sinus and age (p=0.026) as well as a decreasing trend in the presence of an occipital sinus and increasing age (p=0.038). These findings suggest that transverse sinus dominance does vary with age and is more like the adult configuration after five years of age. Codominant transverse sinuses frequency decreased with increasing age, however right transverse sinus dominance increased with increasing age, as seen in other literature (Rollins *et al.,* 2005).

An age-related increase in observed absent transverse sinuses was shown; a transverse sinus was absent in 5% of patients younger than 25 months of age and in 13% of patients 25 months and older. This might be due to alterations in the posterior venous flow patterns, rather than structural involution of the transverse sinus; a decreased volume of blood flow through the transverse sinus may render it a patent sinus which is visible on 2D MRV (Rollins *et al.,* 2005).

Widjaja & Griffiths, (2004) found the right transverse sinus to be dominant in 27 (54%) of 50 patients; left dominant in 18 (36%), and codominant in four patients. One of the patients had absence of both transverse sinuses. A unilateral hypoplastic sigmoid sinus was seen in one patient; 18 patients of the 41 with absent occipital sinuses had hypoplasia of one of their transverse sinuses; three of the 41 had absence of the non-dominant transverse sinus. More cases of hypoplastic or absent transverse sinuses were observed in patients with occipital sinuses than in those without an occipital sinus (67% versus 51%).

Superior sagittal sinus

The SSS is formed from the marginal sinus of each side during embryonic development. The marginal sinus becomes approximated towards the midline as it forms the superior sagittal plexus. However, one should not mistake this marginal sinus with that of the marginal sinus that surrounds the foramen magnum in the adult (Knapp & Schmideck, 1984).

The SSS may remain separated into two limbs (posteriorly not fusing at the midline), each of which drain laterally into their respective transverse sinuses on each side. The two limbs usually join to form the confluence of sinuses during the sixth fetal month. Connecting vascular channels of networks may be found posteriorly where the two limbs of the SSS have not fused. These networks may produce many variations in the region of the confluence of sinuses and adjacent dural structures in later life (Widjaja & Griffiths, 2004).

A single pial cortical vein on each side may be seen if the SSS fails to fuse anteriorly, the veins running posteriorly on each side on the superior aspect of the frontal gyrus in the parasagittal plane, about 1 - 3 cm from the midline. Opening into the SSS behind the coronal suture are the frontal cortical veins. This anatomic variant was observed in one of the cases of Widjaja & Griffiths, (2004).

In the study by Widjaja & Griffiths, (2004) of 50 children aged three months to 17 years, they noted that variations of the SSS included high splitting of the SSS in two patients and a foreshortened SSS, drained by two prominent cortical veins anteriorly in one patient.

Vein of Galen

The vein of Galen reaches its adult configuration during the sixth fetal month with the development of the splenium of the corpus callosum (Knapp & Schmideck, 1984).

Two of the patients in the study by Widjaja & Griffiths, (2004) had a bulbous prominence of the vein of Galen which was not accompanied by a fistula. It is thought that this bulbous prominence is the remnant of the falcine sinus. This is a normal variant in an asymptomatic patient, as it should not be mistakenly interpreted as an aneurysm of the vein of Galen (Widjaja & Griffiths, 2004).

Straight sinus

During embryonic development the cerebral hemispheres enlarge, causing the SS to be displaced and flattened downward (Knapp & Schmideck, 1984). An accessory SS, known as the falcine sinus temporarily appears during the fifth month of gestation. The falcine sinus appears in the falx cerebri just above the proper SS and connect the vein of Galen with the SSS. The SS increases in calibre in the sixth month of gestation, with the course of the sinus declining with the descent of its the torcular portion (Yokota *et al.*, 1978).

Inferior sagittal sinus

Despite a thorough search of the literature, no information could be found on the antenatal development of the inferior sagittal sinus.

Occipital sinus

The occipital sinus is contained within the leaves of dura and connects the confluence with the internal jugular vein. The sinus may be solitary, duplicated or composed of a mesh of venous collaterals (Ruiz *et al.,* 2002) It is thought that the occipital venous network becomes involuted after most of the venous flow passes through the large dural sinuses as the child achieves an upright position (Okudera *et al.,* 1994).

Widjaja & Griffiths, (2004) found that the occipital sinus was absent in 41 of 50 patients. Nine of the patients had an occipital sinus, of which five were bilateral and four were unilateral. Five of the nine patients with an occipital sinus were younger than 2 years, however patients younger than two years accounted for only 24% (12 of 50) of all cases in the study. Six of the nine patients with an occipital sinus had absent or hypoplastic transverse sinuses and three had normal transverse sinuses. The difference in age between those without and those with an occipital sinus was found to be statistically significant (p=0.01), however the size of the occipital sinus did not seem to correlate with age of the patient.

As a child learns to walk upright, the increased venous drainage to the vertebral venous plexus and the diminution of flow into the internal jugular vein may account for the reduction in size and the number of occipital sinuses in older children and adults (Widjaja & Griffiths, 2004). An age-related regression in the occipital sinuses may occur. In the MRV study by Rollins *et al.,* (2005) the occipital sinus was found in 13% of patients who were 25 months of age but in only 2% of children older than 5 years. One may conclude that an age-related regression in the occipital sinuses occurs. A chi-square trend analysis was conducted by Rollins *et al.,* 2005 and a decreasing trend in the presence of an occipital sinus with increasing age was seen (p=0.038).

In the study by Dora & Zileli, (1980), they found that the presence of the occipital sinus is related to the size of the lateral sinus; however in the study by Widjaja & Griffiths, (2004) it was found that the size of the transverse sinus (more than the sigmoid sinus) is more closely related to the presence of the occipital sinus.

The possibility does exist that the occipital sinuses are not atretic, rather present but collapsed and flow within the occipital sinus is so slow as to be undetected by 2D TOF techniques (Widjaja & Griffiths, 2004).

Internal jugular vein:

Rollins *et al.,* (2005) described narrowing of the sigmoid-IJV junction in 23 patients; and a narrowed midcervical IJV was seen in 12 out of 19 patients in whom the mid-cervical region was imaged on 2D MRV.

For a patient in a supine position, most of the venous drainage of the brain is into the internal jugular veins via the transverse and sigmoid sinuses. On cast corrosion studies of adult cadavers, one can see collateral venous drainage between the posterior cranial fossa and the vertebral system is provided by the anterior, posterior and lateral condylar veins as well as the anterior condylar confluence (Ruiz *et al.,* 2002). In the presence of stenosis of occlusion of the internal jugular veins, drainage from the sigmoid sinus may be provided by the posterior condylar vein, which originates from the posterior-superior surface of the sigmoid-IJC junction and drains into the deep cervical plexus (Rollins *et al.,* 2005).

2.8.2 Cerebral venous system of the adult

One may define venous drainage dominance of the dural sinuses as drainage only or mainly to one of the transverse/sigmoid sinuses. The various connections between the sinuses that contribute to the formation of the confluence of sinuses and their accompanying variations has a significant effect on venous drainage dominance. (Durgun *et al.*, 1993).

Fundamental to the disciplines of neurosurgery and radiology are an understanding of the cerebral venous sinus anatomy. It is especially important in surgical planning and treatment of neurological diseases. One should consider dominance in the pattern of cerebral venous drainage in patients undergoing neurosurgery for a large variety of neurological diseases; especially when coagulation of venous structures is possibly required (Durgun *et al.*, 1993; Kitamura *et al.*, 2017). A venous sinus analysis by angiography is recommended for preoperative evaluation of disease of the major sinuses (Kitamura *et al.*, 2017).

Transverse sinuses

Normal variation of the cerebral dural sinuses should be identified and understood as not to mistake it for pathologic abnormalities. Hypoplasia, aplasia and flow gaps of the transverse sinuses have been reported on in various literature and should not be mistaken for thrombosis (Sharma & Sharma, 2012).

<u>Summary of dominance in the cerebral venous system</u>

For a summary of the dominance in the cerebral venous system as described in the literature, see table 2.2 below.

Table 2. 2 A summary of the dominance in the cerebral venous system as described in the literature.

Author	Year	Sample size	Study type	Findings
Bayaroğullari *et al.*	2018	211	Magnetic Resonance Venography	Absent right transverse sinus in 4,73% Hypoplastic right transverse sinus in 17,54% Absent left transverse sinus in 12,32% Hypoplastic left transverse sinus in 49,09%
Kitamura *et al.*	2017	100	Cerebral Angiography	Right sided dominance in 36% Left sided dominance in 14% Equal dominance in 50%
Goyal *et al.*	2016	1654	Magnetic Resonance Venography	Equal dominance in 66,9% Hypoplastic left transverse sinus in 21,3% Aplastic left transverse sinus in 4,1% Hypoplastic right transverse sinus in 5,5% Aplastic right transverse sinus in 0,7% Bilateral hypoplastic transverse sinuses in 1,6%

Author	Year	Sample size	Study type	Findings
Sharma & Sharma	2012	100	Magnetic Resonance Venography	Right sided dominance in 73% Left sided dominance in 18% Equal dominance in 9%
Pallewatte *et al.*	2016	unknown	Magnetic Resonance Venography	Right sided dominance in 58% Left sided dominance in 27% Equal dominance in 15%
Friedmann *et al.*	2011	68	Computed Tomography	Right sided dominance in 70 - 80 % Venous system larger in males than females
Manara *et al.*	2010	102	Magnetic Resonance Venography	Right sided dominance in 61% Left sided dominance in 17%
Park *et al.*	2008	31	Cadaver	Right sided dominance in 35.5% Equal dominance in 58.1%
Surendrababu *et al.*	2006	100	Magnetic Resonance Angiography	Right sided dominance in 59% Left dominance in 30% Equal dominance in 10% Aplastic left transverse sinus in 1% Hypoplastic right transverse sinus in 13% Hypoplastic left transverse sinus in 35%
Alper *et al.*	2004	105	Magnetic Resonance Venography	Equal dominance in 31% Hypoplastic left transverse sinus in 39% Aplastic left transverse sinus in 20% Hypoplastic right transverse sinus in 6% Aplastic right transverse sinus in 4%
Ayanzen *et al.*	2000	100	Magnetic Resonance Venography	Right sided dominance in 59% Left sided dominance in 25% Equal dominance in 16%
Durgan *et al.*	1993	189	Cerebral Angiography	Right sided dominance in 41,3% Equal drainage in 37,6% Left sided dominance in 18,5% Drainage to right side only in 2,1% Drainage to left side only in 0,53%
Browning	1953	100	Post-mortem examination	Right sided dominance in 51% Left sided dominance in 29% Equal dominance in 20%
Gibbs *et al.*	1934	25	Cadaver	Right dominance in 52% Left dominance in 24% Equal dominance in 24%

Author	Year	Sample size	Study type	Findings
Edwards	1931	25	Cadaver	Right dominance in 50% Left dominance in 42% Equal dominance in 8%

The transverse sinus on average measured 6.4 mm on the right and 5.4 mm on the left in the study done by Kitamura *et al.*, (2017). The study found no association between dominance in the transverse sinus and sex.

Manara *et al.*, (2010), investigated a sample of 102 patients. In their samples the right transverse sinus had a significantly greater mean diameter than the left. Manara *et al.*, (2010) did not find any correlation between age and dominance of the transverse sinuses. Among the 204 transverse sinuses evaluated 86 presented with flow gaps.

A small calibre of a transverse sinus has other important implications. Even in patients without variations of the transverse sinus, 2D MRV may result in overestimation of the prevalence of transverse sinus atresia (Manara *et al.*, 2010). Alper *et al.*, (2004) showed that 2D MRV revealed transverse sinus atresia in about 20% of left transverse sinuses and 4% of right transverse sinuses. Durgun *et al.*, (1993) found the cumulative prevalence of aplasia was 2.63%.

Kitamura *et al.*, 2017 found that there was no association between age and sex and the pattern of dominance of the cranial venous sinuses; however, there was a significant association between the SSS and the dominance of the transverse sinus, sigmoid sinus and internal jugular vein.

Durgun *et al.*, 1993 showed that in most cases when the confluence of sinuses is symmetrical the drainage patterns were equally dominant to the left and right transverse sinuses. They also showed that when there was left or right dominance the drainage type of the confluence showed deviation to the dominant side.

One should not regard the confluence of sinuses as purely the drainage site of the superior sagittal and SS or just the origin of the transverse sinuses; one should rather consider it as a complex system of anastomosing sinuses. Therefore, the variations of dural venous sinus drainage into the confluence is an important variable that determines venous drainage dominance (Durgun *et al.*, 1993).

Due to the small samples of studies done by Edwards (1931) and Gibbs *et al.*, (1934) it is doubtful how statistically significant their findings are.

One can see that an understanding of the venous drainage dominance is of particular importance for patients undergoing a radical neck dissection. Removal of neck tumours that invade the internal jugular vein or ligation of the internal jugular vein on the dominant side due to tumours of the glomus jugulare may have important complications. Thus, one may conclude that these surgical complications may be higher in patients with either a right or left sided dominant drainage pattern (Durgun *et al.*, 1993).

In some cases, sufficient venous drainage is provided through the vertebral venous plexus after bilateral removal of the internal jugular veins (Barber *et al.*, 1961). However, the study done by Morfit *et al.*, (1958), showed that after bilateral removal of the internal jugular veins, patients are more prone to postoperative complications than compared with unilateral removals.

Some authors support that adaptation to a new venous drainage type depends heavily on the variations of the venous system and ligation of the dominant versus the non-dominant side (Jones, 1951; Morfit *et al.*, 1958; Kaplan *et al.*, 1972).

Sigmoid sinuses

Goyal *et al.*, 2016 found the sigmoid sinus to be bilaterally present in 1418 of 1654 (85.7%) cases. The left sigmoid sinus was hypoplastic or aplastic in 189 (11.4%) of cases, while the right sigmoid sinus was hypoplastic in 40 (2.4%) of cases and aplastic in 4 (0.2%) of cases. Only three patients had bilateral hypoplastic sigmoid sinuses. Symmetrical sigmoid sinuses were more common in female patients (87.2% in females versus 83% in males, p=0.02). The differences in other variations of the sigmoid sinus was not found to be statistically significant between males and females.

During an autopsy study conducted by Goto, (1986) they found the right sigmoid sinus to be dominant in 70.2% of cases, equal in 18.3% and left dominant in 11.5% of cases. In the study by Kitamura *et al.*, 2017 the sigmoid sinus was right dominant in 35% of cases, left dominant in 15% and symmetrical in 50% of cases.

In the study by Bayaroğullari *et al.*, (2018), the right sigmoid sinus was not seen in 7 (3.31%) of cases and hypoplastic in 34 (16.11%). The left sigmoid sinus was absent in 2 (0.94%) cases and hypoplastic in 72 (34.12%) of cases; among these cases 14 (6.63%) had a bilateral hypoplastic transverse sinus. An occipital sinus was seen as an alternative pathway in cases where the transverse or sigmoid sinus was hypoplastic; this was seen in 30 cases. Two of these cases had a double occipital sinus.

Surendrababu & Livingstone, 2006 found the sigmoid sinus was hypoplastic on the right side in six patients and in 19 patients on the left, with two patients having an aplastic sinus on the left. In 39% of patients the SSS drained to the right transverse sinus; 15% drained to the left and the remaining into the confluence. The SS, however, drained mostly into the left transverse sinus (21%) and only drained into the right transverse sinus in 13% of patients, with the remaining draining into the confluence.

Superior sagittal sinus

The SSS was observed in all cases studied by Ayanzen, *et al.* in 2000.

In a study done by Kitamura *et al.*, 2017, a total of 104 cerebral angiograms were retrospectively obtained from two reference centres. Four angiograms were excluded due to venous sinus thrombosis, and 100 angiograms were included in the study sample. The study found that the division of the SSS was parasagittal to the right in 34% of cases, to the left in 12 % and sagittal in 54% of cases. The study found no statistically significant association between sex or age and venous sinus dominance. The SSS may open directly into the transverse sinus; in cases such as this the opposite transverse sinus is generally hypoplastic or absent. One may also find that the SSS continues as the right transverse sinus and the SS forms the left transverse sinus; this is associated with hypoplasia of the confluence of sinuses. A septum may be present within the confluence having implications in the analysis of the flow within the confluence (Kitamura *et al.*, 2017).

Kitamura *et al.*, 2017; noted that the SSS often deviates slightly to the right, as it courses inferiorly along the occipital bone; in these cases, the SSS ends by forming the right transverse sinus. The study also noted that in about 20% of 'normal' patients the SSS deviates more than 1 cm from the midline. In this study it was also found that the SSS runs sagittal in 54% of the cases, parasagittal right in 34% and parasagittal left in 12% of the cases.

According to Kitamura *et al.*, 2017 the presence of a parasagittal right division of the SSS was associated with the dominance of the right transverse sinus, sigmoid sinus and the internal jugular vein. Presence of a left parasagittal division of the SSS was associated with left dominant transverse sinus, sigmoid sinus and internal jugular vein.

In the study by Goyal *et al.*, (2016) 1616 of 1654 (97.7%) patients had a normal SSS. They found the most common variation of the SSS to be atresia of the anterior one third, other variations seen included, hypoplasia of the anterior one third, hypoplasia of the anterior two thirds and hypoplasia of the anterior half. No statistically significant difference was noted between males and females.

A hypoplastic rostral SSS was report in seven out of 382 cases in one series of anatomic specimens and in 12 out of 201 cases for a second anatomic series by Kaplan & Browder (1972). A large pair of parasagittal superior frontal cortical veins can be seen in cases where the rostral part of the SSS was completely hypoplastic. The parasagittal superior frontal cortical veins run dorsally to join the origin of the SSS close to the coronal suture to replace the absent rostral part of the SSS (Kaplan & Browder, 1973). In a study of 100 patients by CT angiography, complete hypoplastic rostral SSS was reported in 3% of cases (Kaplan & Browder, 1973). Surendrababu e*t al.*, (2006) found this abnormality in 9% of cases in their study. Goyal *et al.*, 2016; found atresia of SSS in 0.9% of cases and hypoplasia in 0.4%.

Sharma & Sharma, (2012) reviewed 100 MRVs from patients with normal MR examinations of the brain parenchyma; all patients had a visible SSS. The SSS was also always seen in the study by Pallewatte *et al.*, in 2016. Hypoplasia was noted in 5.7% of cases, with the anterior quarter being hypoplastic in 3.3% of cases and the anterior half in 2.4% of cases (Pallewatte *et al.*, 2016).

Straight sinus

In the study by Ayanzen, (2000), in which the author reviewed 100 MRVs, the SS was present in all cases. In contrast to this Goyal *et al.*, (2016) conducted a review of 1654 patients who underwent MRV and found the SS to be hypoplastic in four (0.2%) of the cases in males and in three (0.3%) of the cases in females. They found that the difference between the sexes was not statistically significant (p=0.92). Goyal *et al.*, (2016) found no duplication or triplication of the SS in the study population.

Of the 100 MRVs reviewed by Sharma & Sharma (2012), the SS was seen in 91 cases and incompletely visualised in only nine cases.

Inferior sagittal sinus

The inferior sagittal sinus (ISS) was only observed in 52% of the 100 cases studied by Ayanzen *et. al.* (2000). Bayaroğullari *et al.,* 2018 observed the ISS in 91 out of 211 (43.12%) cases and not seen in 120 (56.87%) of cases. Sharma & Sharma, 2012, observed the ISS in only 11% of the 100 cases in their study.

Occipital sinus

Ayanzen *et al.* (2000) observed an occipital sinus in 10 out of their 100 the cases, while Kitamura *et al.* (2017) found the occipital sinus present in 18 of 100 patients; while the right mastoid vein was present in 29% and the left only in 22% of the 100 cases. The suboccipital venous plexus was found in 58% of cases. The occipital sinus acted as a major drainage channel to the SS or SSS in "about 2%" of the cases (Kitamura *et al.,* 2017).

Goyal *et al.,* (2016) identified the occipital sinus in 23 of 1654 (1.4%) of cases; of these five were men and 18 women. The difference between the sexes was not found to be statistically significant (p=0.25). A persistent occipital sinus was seen in 13% of patients less than 25 months of age in a study of 100 children; however, in children older than five years, only 2% had a persistent occipital sinus (Rollins *et al.,* 2005).

Surendrabadu & Livingstone, (2006) identified the occipital sinus in 17 (of 100) patients, while Sharma & Sharma (2012) identified the sinus in four out of 100 patients in their sample.

Over reporting of the incidence of the occipital sinus in the literature may be due to inaccurate interpretation of other venous structures as the occipital sinus or ethnic differences of the populations studied (Goyal *et al.,* 2016).

Petrosquamous sinus

Generally, the petrosquamous sinus is a fetal vein which disappears by birth. When present it runs backwards along the junction of the squamous and petrous portions of the temporal bone and opens into the transverse sinus (AnatomyNext, Petrosquamous sinus, retrieved on 26 December 2019). Anteriorly, the petrosquamous sinus connects with the retromandibular vein through a postglenoid or squamous foramen (Standring, 2008).

Vein of Galen

The vein of Galen was observed in all 100 cases in the study by Ayanzen, *et al.* (2000). Sharma & Sharma, (2012) found the vein of Galen to be present in 96 of 100 cases in their study.

Internal cerebral veins

The paired internal cerebral veins were seen in all 100 cases by Ayanzen, *et al.* (2000). Sharma & Sharma, (2012) identified internal cerebral veins in 60 (of 100) cases.

Basal vein of Rosenthal

The paired basal veins of Rosenthal were seen in 91% of the 100 cases studied by Ayanzen, *et al.* (2000).

Bayaroğullari *et al.*, (2018), identified the vein of Rosenthal in 183 of 211 (86.72%) cases and it was not seen in 28 (13.27%) of cases. The vein was seen in 34 (of 100) patients in the study by Sharma & Sharma, (2012).

Vein of Trolard

The right and left veins of Trolard was observed in 37% and 34% of cases respectively in the study by Ayanzen, *et al.* (2000). The vein of Trolard was seen in 198 of 211 (90.74%) cases on the right, whereas in 13 cases it was not seen on the right. It was seen in 201 of 211 (90.74%) cases on the left and not seen in 10 cases on the left (4.73%) (Bayaroğullari *et al.*, 2018). The vein of Trolard was identified as a relatively constant vein in the study by Surendrababu & Livingstone (2006), as it was present in 79 patients on the right and in 75 on the left out of 100 cases.

Vein of Labbé

The right and left anastomotic vein of Labbé was seen in 91% and 96% of cases respectively (Ayanzen, *et al.*; 2000). Bayaroğullari *et al.*, (2018) observed the right vein of Labbé in 172 (81.51%) of their 211 patients, it was not seen in 39 (18.48%) patients; the left vein was observed in 183 (86.72%) and not seen in 28 (13.27%) of patients. Surendrababu & Livingstone, (2006), found the vein of Labbé in 70 (of 100) patients on the right and in 68 (of 100) on the left.

Internal jugular veins

The internal jugular vein is a continuation of the sigmoid sinus; the initial portion has a small expansion called the jugular bulb. The vein ends by uniting with the subclavian vein to form the brachiocephalic vein. The autopsy study by Goto (1986), showed the internal jugular vein to be right dominant in 67.4% of cases, left dominant in 12.1 % and equal in 20.5% of cases. Kitamura *et al.*, 2017 showed symmetry of the sigmoid sinuses in 56% of cases, 33% were right dominant and 11% of cases were left dominant.

As mentioned previously, the jugular bulb is the initial portion of the internal jugular vein in the jugular fossa. Occasionally, the upper portion is located higher than normal, usually unilaterally. This could be the cause of pulsatile tinnitus due to its close relationship with the pneumatised petrous bone. Knowledge of a higher jugular bulb has value when planning surgeries through the temporal bone.

Friedmann *et al.*, (2011) showed the jugular bulbs were not detected in patients younger than two years; they became enlarged in adulthood and remained stable in the elderly. Kitamura *et al.*, 2017 found that in 70% of cases the jugular bulb was located at the same level on both sides; 19 patients had a higher jugular bulb on the right and 11 had a higher bulb on the left.

<u>2.9 Sex differences in cerebral venous drainage patterns</u>

There is limited data available on the sex differences of intracranial vasculature, even though many neurological and psychiatric conditions differ between males and females with regards to their prevalence and symptomatology (Goyal *et al.*, 2016).

In the meta-analysis study done by Ruigrok *et al.*, (2014) on the sex difference in human brain morphology; the difference in overall brain volumes between male and female was found to be consistent from newborns to persons over 80 years old. On average, males have larger intracranial volume (by 12%), total brain volume (11%), cerebrum (10%), grey matter (9%), white matter (13%), cerebrospinal fluid (11.5%) and cerebellum (9%) absolute volumes compared with females (Ruigrok *et al.*, 2014).

In a study of 257 cases in persons aged 5-82 years, the authors studied the anatomical variations of sigmoid sinuses on phase contrast MR angiography. The study included 91 males and 166 females. The frequency of reduction of flow through the right sigmoid sinus was slightly higher in females than in males. The frequency of reduction of flow through the left sigmoid sinus increases with age in the male group, compared to an increase in the frequency of reduction of flow in the right sigmoid sinus with of age for the female group (Savelyeva *et al.*, 2012). Goyal *et al.*, 2016 found that females were more likely to have symmetrical transverse and sigmoid sinuses than males. The male group had more hypoplastic left transverse sinuses compared with the female group. The significance of the sex differences in the intracranial venous sinus anatomy is not known. In contrast to the findings by Goyal *et al.*, (2016) sex differences in the arterial system have been described in a small MR angiography study of 30 cases (Stefani *et al.*, 2013). The study found that males have a larger diameter of the posterior circulation arteries (posterior cerebral and basilar arteries) than females. The study found no significant difference between males and females in the anterior circulating arteries (Stefani *et al.*, 2013).

2.10 Magnetic Resonance Imaging

There is an important place for magnetic resonance venography (MRV) in the study of the anatomy of the intracranial venous system and its variations (Mattle *et al.*, 1991; Ayanzen *et al.*, 2000).

Magnetic resonance venography is often used as a non-invasive means of examining the intracranial venous system (Ayanzen *et al.*, 2000; Bono *et al.*, 2003). The use of MRV in this setting is increasing in frequency. It is particularly helpful in the diagnosis of venous sinus thrombosis (Ayanzen, *et al.*, 2000).

The most widely used technique is probably 2D time-of-flight (TOF), although other sequences are available. The popularity of this technique is due to the rapid time it takes to complete, and the venous structures may be visualised without contrast (Rollins *et al.*, 2005).

2.10.1 Magnetic resonance venography (MRV)

Visualisation of the venous system can also be done by other methods, for example CT venograms and conventional catheter venography; however, MRV has many benefits (Farb *et al.*, 2003; Widjaja & Griffiths, 2004). One can evaluate the pathology in the vessels close to the bone surface which are difficult to visualise with CT angiography due to the artefacts (Wetzel *et al.*, 1999).

There are a variety of options available for imaging and evaluation of the dural venous sinuses and cerebral veins; this includes digital subtracted angiography (DSA) and MRV (Yigit *et al.*, 2012).

Conventional angiography has also been used to image the venous system (Wetzel *et al.*, 1999; Widjaja & Griffiths, 2004); however, MRV is increasingly being used to visualise and study the cerebral venous system. It is a non-invasive technique, not requiring ionizing radiation and has relatively short acquisition time (Widjaja & Griffiths, 2004).

The cerebral venous sinuses may be visualised and evaluated using either un-enhanced MRV techniques, such as 2D or 3D TOF MRV, phase-contrast MRV and phase-sensitive imaging MRV or contrast-enhanced MRV sequences (Yigit *et al.*, 2012).

Conventional and CT angiography uses ionizing radiation and has an inherent invasiveness to the procedures which is a great disadvantage. Thus, MRVs has a distinct advantage over other modalities, as it can be acquired with or without contrast media (Sharma & Sharma, 2012).

There are two techniques that may be used to visualise the venous anatomy with MRV: TOF and phase-contrast (PC) techniques. Time of flight techniques differentiate moving blood from stationary tissues by the excitation of blood in one slice of the body and detecting the excitation of blood in another slice. Phase-contrast techniques on the other hand, depends on the phase change of moving protons in the blood, this is done by subtracting data acquired twice using positive and negative gradients (Pallewatte *et al.*, 2016).

Disadvantages of phase contrast techniques include relatively long acquisition times and the need to match the imaging gradient sensitivity to the anticipated flow conditions based on deductions (Surendrababu & Livingstone, 2006).

The ability to visualise the intracranial venous system for certain clinical situations such as the diagnosis of cortical venous thrombosis and assessment of the patency of the venous sinuses if encased by meningiomas is important for proper diagnosis and treatment (Goyal *et al.*, 2016).

2.10.2 Contrast enhanced imaging

The superiority of contrast-enhanced MRV techniques to un-enhanced techniques has been demonstrated by various authors (Meckel *et al.*, 2008 ; Lettau *et al.*, 2009; Fu *et al.*, 2010 ; Sun *et al.*, 2010 ; Yigit *et al.*, 2012).

The study by Tapia *et al.*, (2014) showed that CT angiography (contrast enhanced) or MR angiography without contrast may visualise medium sized dural sinuses; however, contrast enhanced imaging is necessary to exclude small cortical venous thrombosis and to determine flow and patency in small sinuses.

Three-dimensional contrast-enhanced gradient-echo techniques are more sensitive to dural sinus thrombosis when compared with 2D TOF MRVs. The 3D contrast-enhanced technique is less affected by saturation effects and they enable better differentiation or atretic and hypoplastic sinuses (Liang *et al.*, 2002).

2.10.2.1 Gadolinium

Magnetic resonance imaging usually uses a gadolinium contrast medium, which is a chemical substance. Gadolinium is injected intravenously and enhances the quality of the MRI images, which allows more accurate reporting of pathology and anatomy. The contrast media consists of complex molecules held together by chemical bonds. These bonds are made between the gadolinium ion and a carrier molecule/chelating agent. The chelating agent prevents toxicity while maintaining the contrast properties of the gadolinium (Ferris & Goergen, 2018).

Functionality of blood vessels in real time may be visualised using gadolinium contrast medium; overall it improves the diagnostic accuracy in many conditions, such as inflammatory and infectious disease of the brain, spine, soft tissues and bones. One may also assess the nature and extent of some cancers and benign tumours using gadolinium contrast (Ferris & Goergen, 2018).

2.10.3 Imaging in the fetus

The development of high-resolution ultrasonography with color-coded Doppler imaging offered a breakthrough in the evaluation of the human fetal venous system. This evolution of technology considerably enhanced the understanding of the fetal venous circulation under normal physiological conditions as well as enabling the study of circulation under abnormal circumstances (Fasouliotis *et al.*, 2002). Fetal MRIs may be used to visualise different organs of the fetus, for example: the brain, spine, face & neck, chest & lungs, abdomen & pelvis and also the placenta (Bailey, 2016).

2.10.4 Imaging in children

Previously the majority of MRV studies have been performed in adults, with little information available regarding the sinus and venous anatomy that can be shown on MRV in children. There is considerable data available from conventional angiography regarding intracranial veins and sinuses, however data regarding the capacity of MRV in children are scarce (Widjaja & Griffiths, 2004).

Use of MRV in the paediatric population has received little attention in the literature. However, familiarity with the normal age-related anatomy on 2D TOF MRV is necessary for radiologists to interpret the MRV images in children (Rollins *et al.*, 2005).

Several methods are used to visualise the intracranial sinus and venous anatomy, including the conventional angiography, MRV and CT venography. Magnetic resonance venography has several advantages over conventional angiography and CT venography in children, based on the radiation dose associated with CT and conventional angiography and invasiveness of conventional angiography. For these reasons the intracranial venous system in children is increasingly being studied by using MRV (Widjaja & Griffiths, 2004).

2.10.5 Imaging in adults

Digital subtraction angiography (DSA) has for a long while been considered the method of choice for evaluating the intracranial venous system in adults (Haroun *et al.*, 2005; Manara *et al.*, 2010). Today there are several different imaging techniques available, all with their own advantages and limitations (Manara *et al.*, 2010). However, with the advent of MRI and MRA techniques, and the change in imaging diagnostic capabilities, these techniques have become the modalities of choice to visualise the cerebral venous system for certain authors (Vogl *et al.*,1994; Ozsvath *et al.*, 1997).

The demand to evaluate the intracranial venous system has increased; this ability is essential in planning for neurosurgery in brain tumours which are contiguous to the dural sinuses and also to identify venous sinus thrombosis (Manara *et al.*, 2010).

2.10.6 Limitations of Magnetic Resonance Imaging

MRV also has few disadvantages, namely it requires longer acquisition times, which may be overcome by the newly developed parallel imaging technique and due to the narrow bore size and loud noise some patients may experience claustrophobia, however ear plugs and an open or wide bore may overcome this. Movement of the patient may cause artefacts and signal loss (Goyal *et al.*, 2016).

Due to a potential lack of sensitivity to flow because of the inappropriate selection of the velocity encoding value, phase-contrast venography may be very limited; it is also a relatively lengthy process (Rollins *et al.*, 2005). Furthermore, MRV interpretation may be difficult due to flow artefact and the presence of anatomical variations (Bono *et al.*, 2003).

2.10.6.1 Limitations of imaging in fetuses

Due to the movement of the fetus during the imaging process, it may be difficult to obtain high quality images. It is recommended that mothers do not eat or drink within the four hours prior to a fetal MRI, as this may cause the fetus to move more. A fetal MRI may take between 30 to 45 minutes, longer times are needed to image twins (UCSF, 2017).

2.10.6.2 Limitations of imaging in children

Our understanding of brain development during childhood and adolescence has been significantly increased by quantitative magnetic resonance imaging (MRI). Giedd *et al.*, 1999 and Sowell *et al.*, 2004 have published large scale longitudinal studies. Unfortunately, none of these studies included children under the age of four years. This is a serious gap in our knowledge (functionally and structurally) regarding brain development in the early years of life, especially in the years from birth until 2 years; which may be the most dynamic and important phase in brain development for humans. The brain increases in size dramatically during this time, reaching 80 -90% of adult volume by age two (Pfefferbaum *et al.*, 1994).

Rollins *et al.*, (2005) noted some problems with saturation effects during 2D MRV, especially in infants, this was mostly observed in the transverse sinuses of neonates. The effects could be due to a number of factors such: as slow blood flow, low volume of flow or in-plane flow, which were most pronounced when the 2D MRV was performed in the axial plane. The in-plane flow could be minimised by changing the orientation of the section slab perpendicular to the expected direction of flow. Coronal imaging ensured better visualization of the transverse sinuses as well as greater anatomic coverage. However, there was a loss of signal intensity due to saturation effects in the IJV's when the images were acquired in the coronal plane in some patients (Rollins *et al.*, 2005).

2.10.6.3 Limitations of imaging in adults

To interpret MR images accurately and avoid pitfalls in diagnosis, one must take note of the technical limitations of 2D-TOF imaging itself. These limitations are mainly related to artefacts resulting from intravascular blood flow, in-plane flow and complex blood flow patterns. 2D-TOF is specifically sensitive to slow flow, it nevertheless has a minimum threshold below which sufficient signal from flowing blood cannot be obtained. If a severe signal loss occurs it could result in the production of artificial flow gaps within a vessel (Ayanzen, *et al.*; 2000).

Saturation effect occurring as in-plane flow and the inclusion of substances with short T1 relaxation time (for example methaemoglobin) are two major drawbacks of time-of-flight techniques (Wetzel *et al.,* 1999). Additionally, 2D TOF MRVs are affected by a progressive signal loss which is caused by slow flowing protons and/or by the flow of protons that are parallel rather than perpendicular to the imaging plane. This may result in spin saturation and cancellation (absence) of segments of sinuses on imaging. Thus, a coronal acquisition plane may occasionally fail to show continuity of the transverse sinus (Manara *et al.,* 2010).
In the case of complex flow patterns, for example in flow separation with vortex flux, haemodynamic conditions can occur that are difficult to image and may contribute to signal intravascular signal loss (Ayanzen, *et al.*; 2000).

A disadvantage of CT venography when compared with MR angiography is the radiation exposure to the patient, and it is also more invasive with inherent complications (Wetzel *et al.,* 1999; Surendrababu & Livingstone, 2006).

<u>2.11 Pathology of the cerebral venous system</u>

It is important to have a working knowledge of the anatomical variations of the cerebral dural venous sinuses. Without such knowledge of the variations one may mistakenly interpret flow gaps, aplasia and hypoplasia of the sinuses as venous sinus thromboses (Goyal *et al.,* 2016).

A pre-operative assessment of the dural sinus pattern done by contrast or MRV is recommended to avoid surgical risk, especially in cases of occipital or suboccipital craniotomies (Singh *et al.,* 2004).

Stroke together with ischemic heart disease are the leading causes of death in the world, accounting for 15.2 million deaths in 2016, of which stroke alone caused around 6 million deaths (WHO, 2011). Even though, fewer than 1% of all cerebral strokes are due to cerebral venous and sinus thromboses (CVST); the frequency may be much higher due to the difficulties in establishing a firm diagnosis (Verulashvili, 2011).

Rather than perceiving it as merely a fibrous structure, the dural venous sinuses should be viewed as anatomically significant structures with implications for neurological surgeries and interventions.

2.11.1 Fetus

The venous system of the fetus is often associated with cardiac or other venous malformations. These abnormalities occur sporadically, while the pathophysiologic mechanisms of their development *in utero* remain largely unknown (Fasouliotis *et al.,* 2002).

In a fetus with anomalies of the venous system a prenatal evaluation including a review for cardiac anomalies in necessary. During this evaluation the pulmonary venous drainage, umbilical, portal, hepatic and ductal systems are thoroughly reviewed. These evaluations are aimed at determining deviations from the norm and to discover any systemic disease or thromboembolic events (Fasouliotis *et al.,* 2002).

2.11.2 Child

Infantile subdural haemorrhage (SDH) is often traumatic in origin and ruptured bridging veins are thought to be the source of subdural blood in these circumstances. However, infantile SDH can also occur in a number of non-traumatic conditions. Some of the conditions include vascular malformations, coagulation disorders and dural venous thromboses. The variety of non-traumatic subdural collections challenges the hypothesis that tearing of the bridging veins are the primary cause of SDH collections. This changes the way one looks at the significance of the dural venous sinuses (Mack *et al.*, 2009), refer to figure 2.29 below.

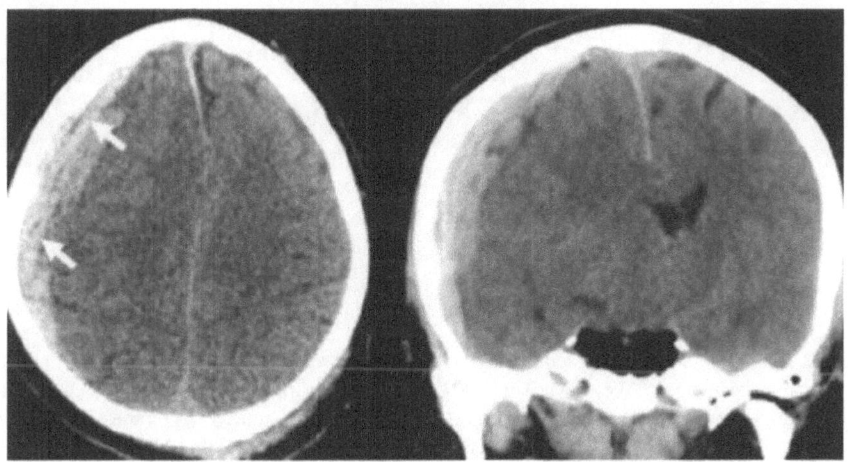

Figure 2.29: Unenhanced CT-scan in the axial and coronal planes showing an acute subdural haematoma (white arrows). From *Grainger & Allison's Diagnostic Radiology, Seventh Edition.* (pg. 1387) A. Adams *et al.,* 2020. Philadelphia: Elsevier. Copyright 2021 by Elsevier. Used with permission.

Premature neonates are commonly affected by germinal matrix haemorrhage-intraventricular haemorrhage (GVM-IVH). This begins in the germinal matrix tissue (the area in the fetal brain that gives raise to the neuronal cells, which will later form the grey matter of the brain) and is most commonly the result of venous rupture. If arteriole-to-venous- precapillary shunts are present in the cerebrum of the neonate, it could lead to elevated venous pressure in the germinal matrix and as a result be an important aetiological factor in GMV-IVH. Under normal circumstances the intravascular pressure falls markedly when blood leaves the arterioles and enters the capillaries. However, when the arterioles are directly connected to the veins by the precapillary shunts or channels, this drop does not occur and the high arterial pressure would pass more or less unchanged into the venous circulation (Anstrom *et al.*, 2002), refer to figure 2.30 below.

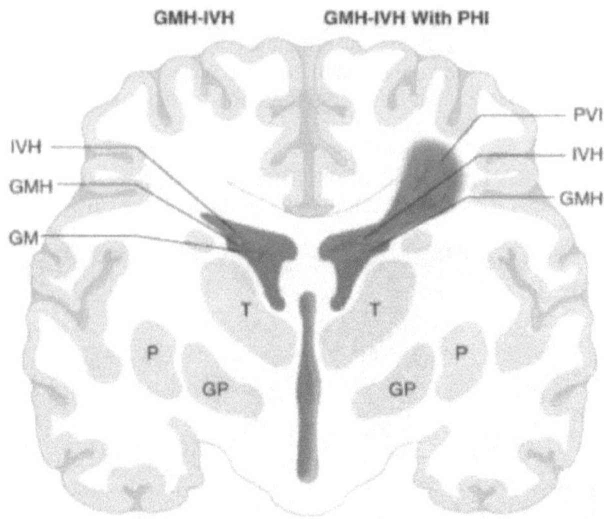

Figure 2.30: Image showing a haemorrhage (red) into the germinal matrix (GM) resulting in germinal matrix haemorrhage (GMH) which could burst through the ependyma to cause an intraventricular heamorrhage (IVH) as seen on the left side of the image. When the GHM-IVH is large, a periventricular hemorrhagic infarction (PVI) may occur (right side of the image). Adapted from Brain injury in premature infants: a complex amalgam of destructive and developmental disturbances, by Volpe J.J, 2009. *The Lancet Neurology,* pages 110-124. Used with permission.

Anstrom *et al.* (2002) evaluated 35 premature neonates obtained at autopsy and concluded that the elevated venous pressure that leads to venous rupture and germinal matrix haemorrhage more likely results from impaired venous flow through the heart and lungs. The impaired venous flow may be caused by different conditions associated with prematurity, for example respiratory distress syndrome and pneumothorax. These conditions may be exacerbated by therapeutic obstetrical manoeuvres that increase thoracic pressure (Anstrom *et al.*, 2002).

2.11.3 Adult

The cerebral veins and sinuses do not have valves, nor do they have a tunica muscularis as part of the vessel wall. Thus, blood flow is possible in different directions. Moreover, the cortical veins are linked by numerous anastomoses, thus allowing the development of a collateral circulation. This could explain the favourable prognosis of some cerebral venous thromboses (Uddin *et al.*, 2006).

The absence of the tunica muscularis permits veins to remain dilated. This is important in understanding the considerable capacity to compensate even with extended occlusion. The cerebral venous sinuses are located between two rigid layers of dura mater, and this prevents compression of the sinuses when the intracranial pressure rises (Uddin *et al.*, 2006).

The superficial cortical veins drain into the SSS against the blood flow in the sinus, thus causing turbulence in the bloodstream, which is further aggravated by the presence of fibrous septa at the inferior angle of the sinus. This explains the greater prevalence of SSS thromboses (Uddin *et al.*, 2006).

The SSS receives blood from the diploic, meningeal and emissary veins, in addition to draining most of the cerebral hemispheres. This is also the case with other dural venous sinuses. This explains the frequent occurrence of CVT as a complication of infective pathologies, for example cavernous sinus thrombosis in facial infections, lateral sinus thrombosis in chronic otitis media and sagittal sinus thrombosis in scalp infections. Dural sinus thrombosis can lead to blockage of the arachnoid villi, which leads to intracranial hypertension and papilledema (Uddin *et al.*, 2006).

2.11.3.1 Sinus pericranii

Sinus pericranii involves congenital or acquired anomalous connections between an extracranial blood-filled nodule and an intracranial dural venous sinus by dilated diploic and/or emissary veins of the skull (Standring, 2008), refer to figure 2.31 below.

Sinus pericranii may be seen as iso- or hyperdense structures on NECT and shows strong uniform enhancement after contrast administration (Osborn, 2013).

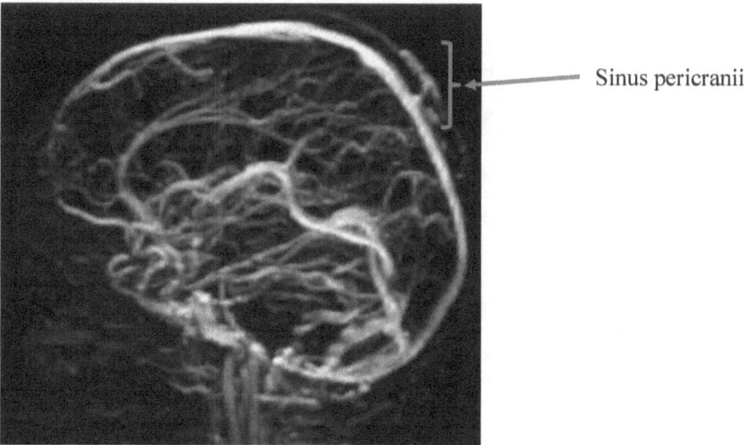

Figure 2.31: Radiologic image showing sinus pericranii (a connection between extracranial blood-filled nodule and an intracranial dural venous sinus, in this case the SSS). Adapted from Sinus pericranii in *Radiopaedia,* by Dr. H. Salam. Retrieved January 04, 2020 from https://radiopaedia.org/cases/13619">rID: 13619</a>. Used with permission.

2.11.3.2 Dural venous sinus thrombosis

Dural venous sinus thrombosis may occur at any age, it has a varied clinical presentation and mode of onset (Bousser, 1992), and it may be difficult to diagnose (Daif *et al.,* 1995). An understanding of the anatomical variations of the venous sinuses and flow gaps will play an important role in preventing overdiagnosis of cerebral venous thrombosis (Surendrababu & Livingstone, 2006). Magnetic Resonance Venography observations of anatomical variations should be taken into account when considering a diagnosis of dural sinus thrombosis (Alper *et al.,* 2004).

Dural venous sinus thrombosis is seen in a number of conditions, including dehydration, hypercoagulable states, infection and tumour invasion; it may also cause neurologic deterioration. Traditionally, the diagnosis is made during the venous phase of conventional catheter angiography. However, this is an invasive procedure with associated risks. Recently it has been reported that MR imaging and MR angiography may be able to replace conventional angiography in the diagnosis of dural venous sinus thrombosis (Ayanzen, *et al.,* 2000).

In the study done by Ayanzen *et al.* in 2000, they observed flow gaps in 31% of patients with non-dominant transverse sinuses whose MR studies were clinically normal. It is thought that these flow gaps, which were only found in the non-dominant transverse sinuses, are artificial in nature. This supposition is supported by the fact that conventional catheter angiography showed the presence of hypoplastic, although patent, non-dominant transverse sinuses, in these cases the corresponding MRVs failed to show sinus continuity. This could potentially lead to diagnostic difficulties when the issue of venous sinus thrombosis is in question (Ayanzen, *et al.,* 2000).

Dural venous sinus thrombosis of the SSS can be identified using the empty delta sign on CT. This triangular filling defect represents the thrombus. It is usually described with CECT-scan (contrast enhanced computed tomography) or MRI, but not with NECT (non-enhanced computed tomography) or non-contrast MRIs. One may also note a similar sign in the transverse sinus when looking in the sagittal or coronal planes (Bell & Gaillard, 2019).

The empty delta sign had been reported in the literature to be present in 29% to 76% of patients with SSS thrombosis or SSS related thrombosis. However, a high splitting SSS may mimic appearance of the delta sign (Wetzel *et al.,* 1999).

The exact mechanism for this sign is still unknown, it might be due to recanalisation around an organising clot, enlargement of the peridural small veins (dural cavernous spaces and meningeal venous tributaries) or due to thickening of the dura with increased enhancement (Bell & Gaillard, 2019).

Dural venous sinus thrombosis may also present with a cord sign (cordlike hyperattenuation) within the dural venous sinus on non-contrast enhanced CTs. This sign is mostly seen in the transverse sinus due to the location of the sinus along the origin of the tentorium cerebelli. One might see a false positive cord sign in the presence of generalised cerebral oedema (Dixon, 2019).

Prominent arachnoid granulations, intrasinus septa and hypoplasia or aplasia of the dural sinuses may mimic venous sinus thrombosis, making neuroradiological diagnosis of venous sinus thrombosis difficult (Manara *et al.,* 2010).

One should be cautious when aiming to achieve a well-timed acquisition with CT cerebral venography, especially in cases of intracranial hypertension that can cause a delay in filling of the venous sinus. A false impression of a thrombosis may be created during a premature acquisition due to the contrast not yet reaching the venous sinuses. One may suspect premature acquisition when the cerebral veins are not opacified on a CT cerebral venogram (Gao *et al.*, 2018).

Cerebral venous thrombosis accounts for only about of 0.5% of all strokes, with the annual incidence of about 5 per million, far less than that of ischaemic and haemorrhagic stroke (Gao *et al.*, 2018).

Younger adults are mostly affected by thrombosis of the cerebral sinus; it may present with a variety of clinical symptoms, from subtle to severe (Martinelli *et al.*, 2010; Saponi *et al.*, 2011). The most common complaint is headache, seen in 71.3% of patients and that may be accompanied by vomiting (Sassi *et al.*, 2017). To reduce the risk of a fatal outcome or disability it is important to diagnose thrombosis early and treat the patient appropriately. Digital subtraction angiography is the gold standard for diagnoses of CVST; however, the procedure is invasive and has radioactive characteristics that may limit its application (Masur *et al.*, 2004).

Venous sinus thrombosis does not necessarily proceed to venous infarction, due to slow progression and the formation of a collateral circulation. However, venous infarction with haemorrhagic transformation may develop when venous sinus thrombosis is associated with cortical or bridging vein thrombosis (Surendrababu & Livingstone, 2006).

In 1925 Tobey and Ayer created a test for the detection of thrombi of the transverse sinus. The test depended on the fact that if the transverse sinus was occluded it will not act as an outflow for the cerebral circulation. Thus, unilateral compression of the vein of the affected side would not produce increased obstruction to the circulation and consequently there would not be a rise in the spinal fluid pressure. The vein on the unaffected side has had to take over the entire outflow, and therefore its compression would cause a rise equal to that of bilateral compression in a normal person.

It is therefore important to understand the configuration of the cerebral venous drainage patterns before attempting an intervention, refer to figure 2.32 below.

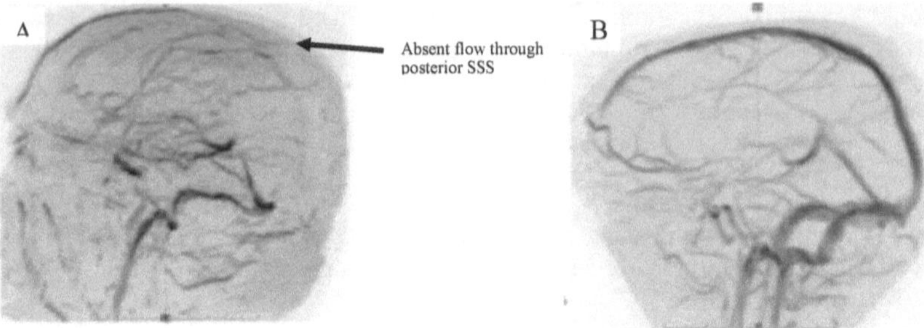

Figure 2.32: A – Magnetic resonance venography demonstrating the absence of flow in the posterior part of the SSS. B – Normal MRV for comparison. Adapted from *Netter's Neurology* (pg.523) by J. Srinivasan *et al.*, 2020, Philadelphia, PA: Elsevier. Used with permission.

2.11.3.3 Vein of Galen aneurysmal formation

A vein of Galen aneurysmal malformation (therefore not a true aneurysm) (VGAM) is the most common cause of high-output cardiac failure in new-borns. A VGAM can be described as a direct arteriovenous fistula between a persistent embryonic precursor of the vein of Galen and the deep choroidal arteries. This malformation causes flow-related aneurysmal dilatation of this primitive vein; this forms a large venous pouch in the midline posterior to the third ventricle (Osborn, 2013).

This type of formation is rare and represents only 1% of all cerebral vascular malformations; it does however, account for 30% of all symptomatic vascular malformations in children (Osborn, 2013).

2.11.3.4 Developmental Venous Anomaly

Developmental venous anomalies or DVA's are umbrella-shaped congenital cerebral malformations, which are composed of angiographically mature venous elements. One would expect to find thin-walled and dilated venous channels within normal brain parenchyma (Osborn, 2013). The exact aetiology is unknown. It is thought that DVA's form due to arrested medullary vein development between 8 and 11 gestational weeks, or it represents a variant of otherwise normal venous drainage (Osborn, 2013). Developmental venous anomalies account for 60% of all cerebrovascular malformation, making it the most common intracranial vascular malformation. The prevalence is estimated at 2.5 - 9% on contrast-enhanced MR imaging (Osborn, 2013).

2.11.3.5 Cerebral cavernous malformation

Cerebral cavernous malformations are characterized by the presence of repeated 'intralesional' haemorrhage into thin walled, angiographically immature, blood filled locules, termed caverns. Cerebral cavernous malformations do not contain any normal brain parenchyma. Cerebral cavernous malformations are commonly the cause of nontraumatic intracranial haemorrhage in young and middle-aged adults; however, they may occur at any age. They present as a well-circumscribed mass of mixed density/signal intensity on imaging, it is surrounded by a complete hemosiderin rim (also known as a popcorn ball) (Osborn, 2013).

2.11.3.6 Cavernous sinus thrombosis/thrombophlebitis

Cavernous sinus thrombosis or thrombophlebitis is a very rare but lethal condition. It is associated with a high morbidity and mortality rate. It can be described as a blood clot in the cavernous sinus, it may be with infection (thrombophlebitis) or without (thrombosis). On a MR scan one can see an enlarged cavernous sinus with convex lateral margins (Osborn, 2013).

2.11.4 Venous occlusion mimics

2.11.4.1 Sinus variants

Normal anatomical variations may account for many diagnoses of cerebral venous thrombosis. Refer to section 2.8 for more information.

2.11.4.2 Flow artifacts

A flow void or gap constitutes a dark or "blank" space on a radiographic image of a fluid-filled structure ("Flow Void," 2020). In the context of an MRI, it is the absence of signal from blood, in which the activated protons leave an area of blood flow before the magnetisation is measured. The term is widely used by radiologist and others involved with MR imaging. One should distinguish that a flow void is not the absence of blood flow, rather the absence/low signal from blood flow through the vessel. The signal loss is usually seen in vessels with vigorously flowing blood and is generally synonymous with vascular patency. Flow voids may also be seen with active flow or pulsation of other fluids, for example CSF or urine (Elster, 2020)

A flow gap in the SSS was identified in five patients by Goyal *et al.*, 2016. Three of these cases had a flow gap in the posterior third of the SSS and two had this in the middle part of the SSS. Surendrababu *et al.* (2006) found flow gaps in 24 % of the SSS and a partial split of the SSS in 12% of patients (n=100) studied. In another study, short segment flow gaps in SSS were observed in 5% of cases (Sharma & Sharma, 2012). Flow gaps were identified in 31% of cases in the transverse sinus in the study by Pallewatte *et al.*, (2016). These flow gaps may be misdiagnosed as cerebral venous thrombi if not evaluated correctly (Pallewatte *et al.*, 2016).

Flow gaps are commonly seen in the transverse sinuses. Ayanzen *et al.*, (2000) found that 90% of the flow gaps they saw were located in the non-dominant transverse sinus; the other 10% were seen in transverse sinuses that were considered codominant, with none being observed in the dominant transverse sinus. In 48% of the cases the length of the flow gap was less than or equal to one third of the length of the ipsilateral sinus. For the remaining cases, 26% had flow gaps greater than one third but smaller than two thirds the length of the ipsilateral sinus and the other 26% had flow gaps larger than two thirds the length of the ipsilateral sinus (Ayanzen *et al.*, 2000).

As mentioned previously, 2D TOF MRA is widely used to visualise the cerebral venous system, however it is very sensitive to slow intravascular blood flow, which may cause an artefact on the imaging known as a flow gap. Flow gaps may also be produced in instances of in-plane flow due to saturation of the MR signal or in complex flow patterns as in the confluence of sinuses (Ayanzen *et al.*, 2000; Yang *et al.*, 2002).

To obtain high signal intensity, the vessel should be perpendicular to the imaging plane. Ideally the intracranial venous system should be imaged in three planes due to its complex 3D structure, however this comes at the expense of time and may not be well tolerated in very ill patients (Haroun, 2005).

It is, however, difficult to distinguish between an aplastic sinus, acute thrombosed sinus or loss of signal within a sinus due to a flow gap artefact with 2D TOF MRA (Haroun, 2005), refer to figure 2.33 below.

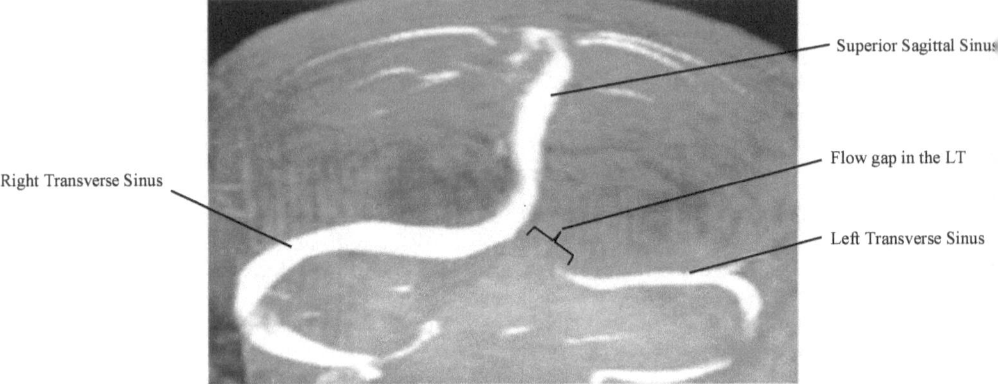

Figure 2.33: Example of a flow gap in the left transverse (LT) sinus. Adapted from Intracranial MRV using Low-Field Magnet: Normal Anatomy and Variations in Nepalese Population, by U.K Sharma & K. Sharma, 2012, *Journal of Nepal Medical Association,* pg.63. Used with permission.

2.11.4.3 Arachnoid granulations and septations

Giant arachnoid granulations may also be mistaken on imaging as thrombi of the SSS. However, one may distinguish between granulations and thrombi as the arachnoid granulations do not fill the entire sinus (as thrombi do when fully formed) and unlike clots, they often demonstrate central linear enhancement (increase in contrast at a specific rate over a given short-time interval) (Osborn, 2013).
Septations may be found in 30% of transverse sinuses, mostly in the right transverse sinus (Osborn, 2013).

2.11.5 Mimics of venous occlusion on images

Several other entities may mimic venous sinus occlusion; these include:
- **High haematocrit** – These may cause the appearance of a hyperdense sinus relative to the brain parenchyma (Osborn, 2013).
- **Unmyelinated brain** - Infants and children commonly have higher haematocrits and a lower density of their unmyelinated brains. The combination of high-attenuated blood vessels and low-attenuated brain causes the vascular structures to appear relatively hyperdense to dural sinus (Osborn, 2013).

- **Diffuse cerebral oedema** - Along with decreased attenuation of the cerebral hemispheres causes the dura and all intracranial vessels to appear hyperdense compared with the low-density brain (Osborn, 2013).
- **Subdural haematoma** - An acute subdural haematoma layer along the SS and medial tentorium cerebelli may cause hyperdensity that may mimic dural sinus thrombosis on NECT scans. This is due to the dense thrombus surrounding the relatively less dense flowing blood in the SSS and in the confluence, mimicking an empty delta sign (Osborn, 2013).

2.11.6 Application in surgical interventions

Patients who undergo radical neck dissections and require ligation of the internal jugular vein should have their venous drainage dominance reviewed as it is of clinical importance to prevent venous drainage impairment of the cerebral structures (Durgun *et al.,* 1993).

The cerebral veins have a highly dynamic nature in response to head positioning and mechanical stimuli which highlight the need for a real-time intraoperative imaging modality which is able to identify dynamic venous modifications (Prada *et al.,* 2018).

Depending on the pathology with which a patient presents, intracranial vascular treatments are minimally invasive procedures that may involve image-guided internal catheter navigation of the blood vessels to treat vascular conditions affecting the intracranial vessels (RSNA, 2017).

2.11.6.1 Application in endovascular therapy

This procedure refers to the precise treatments involving navigation of blood vessels internally with catheters, for example (RSNA, 2017):

- **Embolisation:** Embolic agents (synthetic solid or liquid materials) are placed through a catheter into a blood vessel
- **Drug delivery:** Medications delivered through a catheter to specific locations in the brain through blood vessels
- **Device delivery:** Used to permanently or temporarily insert medical devices through a catheter:
    - Stents: Used to open blood vessels, redirect blood flow and stabilize other devices such as coils
    - Balloons: Used to open blood vessels or assist in placement of other embolic agents or devices
    - Coils: These devices block blood flow, especially used in treatment of intracranial aneurysms
- **Mechanical retrievers/ aspiration systems:** To remove clots, embolic devices or debris

2.11.6.2 Stereotactic radiotherapy

Stereotactic radiosurgery is a form of radiation therapy that uses highly focused X-rays to ablate abnormal blood vessels (RSNA, 2017). Stereotactic radiosurgery may be used to treat arteriovenous malformations (AVM). In an AVM the blood flows directly between arteries and veins, bypassing the capillaries. If left untreated the AVM may cause stroke or lead to bleeding in the brain (Mayo Clinic, 2020). An understanding of the venous drainage pattern and variations is essential when trying to treat these malformations.

2.11.6.3 Application in transtentorial surgeries

Supracerebellar infratentorial approach

The patient is placed in the Concorde position (prone and with the head flexed). From the external occipital protuberance a midline incision is made to the first palpable cervical vertebra. It is not necessary to open the foramen magnum; however, it does facilitate the craniotomy by driving the craniotome in a cephalad direction. One should extend the incision to the confluence of sinuses in order to maximise upward retraction of the dural structures. A U-shaped durotomy is performed based on the transverse sinus (Hart *et al.*, 2013).

An initial inspection is done to ensure all bridging veins are within view. The arachnoid and intervening bridging veins are dissected allowing the cerebellum to fall out of the surgical corridor. Placing a retractor in the paramedian plane will aid in flattening the culmen and holding the brain out of the way. This will also reveal the precentral cerebellar vein (Hart *et al.*, 2013).

The vein of Galen and the internal cerebral vein are inspected microscopically, and the inspection proceeds caudally to expose the tentorium cerebelli and colliculi (Hart *et al.*, 2013).

Occipital transtentorial approach

In the occipital transtentorial approach the patient is prone with their head flexed and rotated slightly towards the side of the approach (three-quarters prone position). The occipital skin flap inferiorly is raised allowing the formation of a free bone flap on both sides of the SSS. The dura is opened as a flap based on the location of the SSS, preserving the bridging veins (Hart *et al.*, 2013).

It is critical to consider the angle of the tentorium cerebelli and its relationship to the deep veins. It is paramount to determine the extent of the tumour in relation with the anatomical blind spots for each approach in order to safely conduct transtentorial surgeries (Hart *et al.*, 2013).

2.11.7 Clinical significance of persisting embryonic venous patterns into adulthood

Anatomical studies of fetal specimens have shown an extensive collateral network connecting the venous sinuses of the posterior fossa with the vertebral and cervical veins (Okudera *et al.*, 1994). By the end of the third trimester of gestation, the internal jugular veins and the sigmoid sinus are connected by the jugular sinus, which remains a small structure and eventually gives rise to the superior jugular bulb Okudera *et al.*, 1994). There is very little information published about the jugular sinus. An increase in venous flow from the rapidly growing cerebral hemispheres leads to ballooning of the transverse sinus, in the absence of an increase in diameter of the sigmoid and jugular sinuses. In early infancy one can see that the diameters of the sigmoid sinuses and internal jugular veins are smaller than the dominant transverse sinuses on MRVs. In the study by Rollins *et al* (2005), the difference in luminal calibre became more pronounced later in infancy and childhood, and it persisted through adolescence. They also found that narrowing of the sigmoid-IJV junction was common.

Any persistence of embryonic venous structures into adulthood will have an impact on surgical approaches and complications. A clear understanding of the normal anatomy and its variations is necessary to facilitate successful treatments of intracranial pathology.

In the embryo there is a continuous network of primitive endothelial channels which drain the brain and dura. These channels undergo venous cleavage which leads to most of the primary venous anastomoses between the pia and dura being resorbed. A few remaining venous anastomoses enlarge to become definitive bridging veins (Mack *et al.*, 2009). It is speculated that if these primitive endothelial channels do not undergo venous cleavage or are not resorbed it could result in persistence of the embryological venous drainage pattern, thus resulting in subdural venous blood pools in the adult. Presence of these blood pools could complicate surgical interventions and might not be seen on normal cerebral imaging.

Chapter 3: Materials and Methods

<u>**3.1 Materials**</u>

3.1.1 Study design

This is a retrospective, descriptive study. Informed consent by the patient was not required due to the retrospective nature of the study. Approval was obtained from the Groote Schuur and Red Cross Children's War Memorial hospitals to access patients record on the condition that no personal identifiers of the patients be collected or kept.

3.1.2 Sample population

The sample population included persons who came to Groote Schuur Hospital for MRIs of the brain; as well as children undergoing MRI of the brain at the Red Cross Children's War Memorial Hospital.

Furthermore, stillborn fetuses from the New Somerset Hospital were included in the sample for the fetal dissections.

3.1.3 Sample selection

3.1.3.1 General considerations

Magnetic resonance images were selected non-randomly from the Departments of Radiology at the Groote Schuur and Red Cross Children's War Memorial Hospitals. The scans were evaluated based on imaging sequence and availability of demographic data. The original indications for the MRIs in these patients were unknown, as the researcher only had access to the MRIs and not the complete patient file, the primary indication for taking the MRI was not known. To maintain confidentiality to the best of the researcher's ability, the permission to view the separate patient file was not requested as a diagnosis or indication was not needed to review the MRIs.

The following structures were used as landmarks and points of reference when describing any variations of the venous drainage pattern and persistence of embryological structures.

Dural venous sinuses:

- Superior sagittal sinus
- Inferior sagittal sinus
- Transverse sinuses
- Sigmoid sinuses
- Occipital sinus

The volume was calculated for the superior, transverse and sigmoid sinuses and the variations in pattern of drainage was described. The inferior sagittal and occipital sinuses were used as landmarks for orientation purposes on the MRI. Since their borders could not be clearly defined on the images their volumes were not calculated, the variations in the patterns of drainage were not described and they were not included in the sample analyses.

Cerebral veins:

- Internal cerebral veins
- Vein of Galen
- Vein of Trolard - Superior anastomotic vein
- Vein of Labbé - Inferior anastomotic vein

3.1.3.2 Fetal dissections

Stillborn full-term fetuses were obtained from the New Somerset Hospital. In total 5 fetuses were dissected. The dissection of the fetuses was to gain a better understanding of antenatal venous structures and they were thus not included in the statistical analyses of the study.

3.1.3.3 Images from children

Magnetic resonance imaging for children aged from birth to twelve years were included in this sample. The MRIs were obtained retrospectively for children who had undergone scans either at the Red Cross Children's War Memorial Hospital or at the Groote Schuur Hospital; images were selected from the picture archiving and communication system (PACS) database.

3.1.3.4 Images from adolescents and adults

Magnetic resonance imaging was obtained retrospectively for persons older than twelve years from the picture archiving and communication system (PACS) database at the Groote Schuur Hospital.

3.1.4 Inclusion and exclusion criteria

3.1.4.1 Inclusion criteria

Images were selected non-randomly.

Scans were evaluated for quality and presence of venous structures. Magnetic resonance images were only selected if the images included the following:

- The entire brain was captured on the scans
- The venous sequence was present
- Post contrast imaging was available for the patient
- Images were captured isometrically (i.e. all slices were the same thickness)

The researcher decided to use post contrast images and not MR venograms (or a combination of post contrast and MRV) for the study as a large enough sample of MRVs could not be obtained. A combination of MRV and post contrast imaging was not used because it has not been established whether data obtained from an MRV (for example volume measurement of the transverse sinus) is directly comparable with the data (also for volume of the transverse sinus) obtained from a post contrast image. All images used in the study were magnetisation prepared rapid gradient echo (MPRAGE) sequence (gradient echo T1 weighted).

3.1.4.2 Exclusion criteria

It is possible, but highly unlikely, that a vessel will be mistakenly marked as absent due to the use of contrast agent and the image being viewed in three separate planes (coronal, sagittal and transverse). For this reason, no images were excluded if any of the vessels were absent, and this would then be reported as a variation in the pattern of venous drainage (aplasia).
Images were excluded for the following reasons:

- Post contrast images were not available because post contrast sequence is required for volume determination)
- Suspicion of venous infiltration by tumours due to the possibility of distortion of the venous volume
- Noted in medical records that previous surgical interventions involving the venous system were performed
- Presence of artefacts obscuring any venous structures (e.g. movement by the patient during imaging)
- Patients where only CT images were available
- Isometric imaging slices could not be obtained. Isometric images are acquired in such a way that they are all of the same thickness, thus if the plane of view is changed the images are not distorted and the same volume will be calculated in all planes for the same structure
- Images of persons with midline shift or any other condition where the aqueduct of the midbrain cannot be visualised, or is occluded as the aqueduct of the midbrain is used to determine the midline of the brain

3.1.5 Sample size

The sample size calculation was based on the following parameters obtained from the Department of Radiology, Groote Schuur Hospital:

- Monthly the staff conduct between 550 and 600 brain MRI procedures
- On average 250 - 300 brain MRIs include contrast per month
- 2 - 3 Magnetic Venograms are done per month

Approximately 5400 scans (300 MRIs with contrast over a period of 18 months) were manually reviewed by the researcher and the inclusion and exclusion criteria were applied.

Data were collected over an 18-month period, from January 2016 to June 2017. During this timeframe and considering the inclusion and exclusion criteria for the study, the sample included 302 patients between the ages of 13 and 85.

Initially the researcher aimed to include children from birth until 12 years of age in the study, however after manual review of all the imaging of the venous sequence, none of the images met the inclusion criteria for the study. It should be noted that MRIs including the venous sequence is not readily available in children and early adolescents (birth to 12 years) due to the long acquisition times and cost involved.

3.1.6 Ethical considerations

Approval for the study was obtained from the following institutions and departments:

- University of Cape Town Ethics Committee (Human Research Ethics Committee)_Ethics reference number: HREC REF:483/2016 (refer to appendix A)
- Head of the Department of Human Biology, University of Cape Town (continuous approval was obtained through the annual progress report submitted to the Department of Human Biology)
- Head of the Department of Radiology at the Groote Schuur Hospital and the Department of Health of the Western Cape (refer to appendix C)
- The Red Cross Children's War Memorial Hospital (The RCCH was notified of the study and acknowledged receipt of the study proposal. Permission was given to the researcher by the Head of the Department of Neurosurgery to obtain credentials for the PACS system. As permission was already obtained to use the facilities at the Groote Schuur Hospital, permission to access the Red Cross facilities was not needed, as the RCCH images could be accessed from GSH)

3.2 Methods

3.2.1 Fetuses

The cerebral vasculature of fresh (unembalmed) fetuses were dissected at the Department of Human Biology of the University of Cape Town. The scalp and skull cap were removed to expose the brain tissue. It proved difficult to obtain a clear view and dissection of the venous structures of the unembalmed fetus brains. Therefore, it was decided to inject the fetuses with a 3% formalin solution and then immerse them in a 5% formalin solution for at least 8 hours prior to dissection. The partial embalming of the fetus hardened the brain tissue enough for clearer dissection, without disturbing the configuration of the cerebral blood vessels.

Method of dissection

An incision was made from ear to ear across the superior aspect of the skull, see figure 3.1 below. The skin was then carefully dissected anteriorly and posteriorly exposing the calvaria and fontanelles. The calvaria was removed by using the anterior fontanelle as entry point. As the calvaria was still soft and flexible and therefore it was possible to cut through the bone with a pair of scissors.

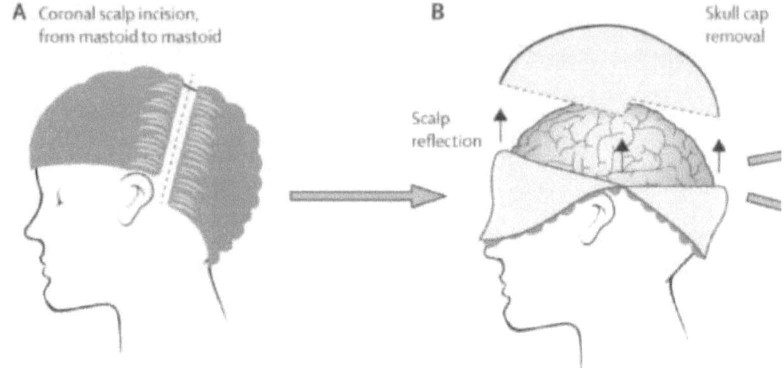

Figure 3.1: Removal of the scalp. From Brain banking for neurological disorders, by Samarasekera *et al.*, 2013. *The Lancet Neurology,* pg. 1096. Used with permission.

The skull was cut along the midline from the anterior fontanelle towards the occipital protuberance and the nasofrontal suture respectively; leaving the SSS *in situ,* to be reviewed and described, see figures 3.2 and 3.3 below.

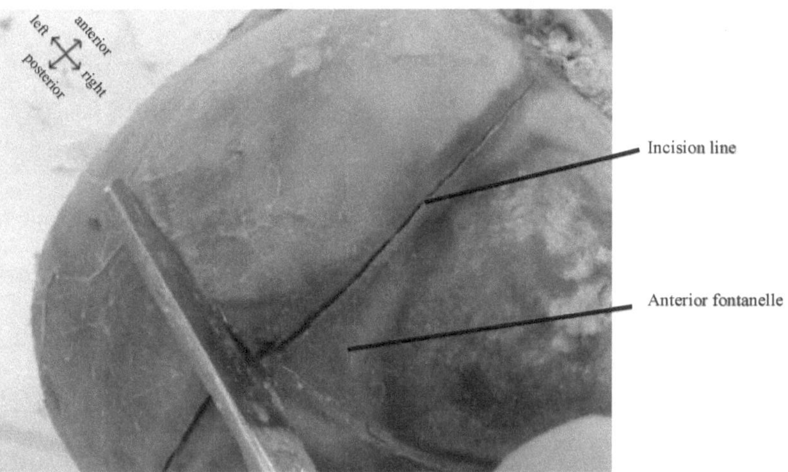

Figure 3.2: Removal of the calvaria. Image from the current research project.

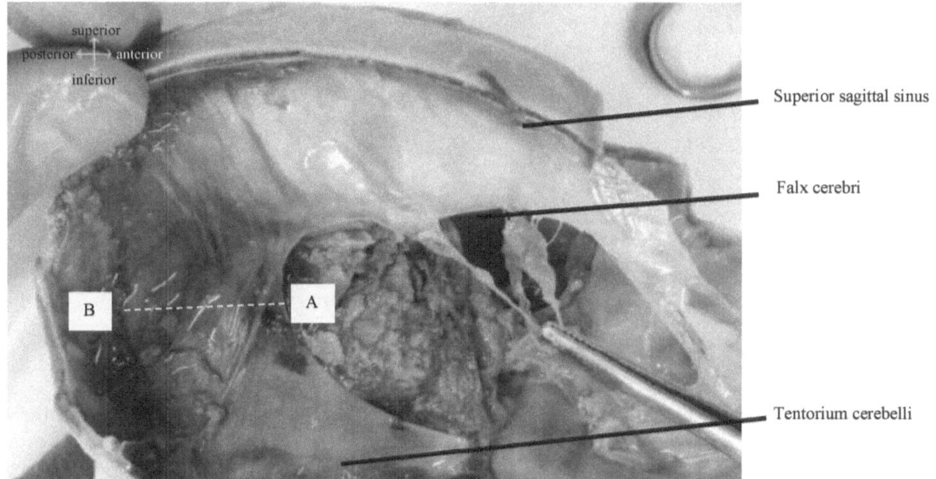

Figure 3.3: Image showing removed calvaria with SSS *in situ,* with A – anterior border of the falx cerebri and B – internal occipital protuberance. Image from the current research project.

The brain was removed using blunt dissection in order to expose the underlying tentorium cerebelli, straight and the transverse sinuses, figure 3.3. The venous sinuses were visually inspected and compared with findings of other studies in the existing literature on the dural venous sinuses antenatally. The falx cerebri was cut from the anterior border (point A) towards the internal occipital protuberance posteriorly (point B) to better visualise the confluence of sinuses and the transverse sinuses, figure 3.3.

The transverse sinus was opened and the drainage pattern into the confluence of sinuses evaluated, figure 3.4 and 3.5.

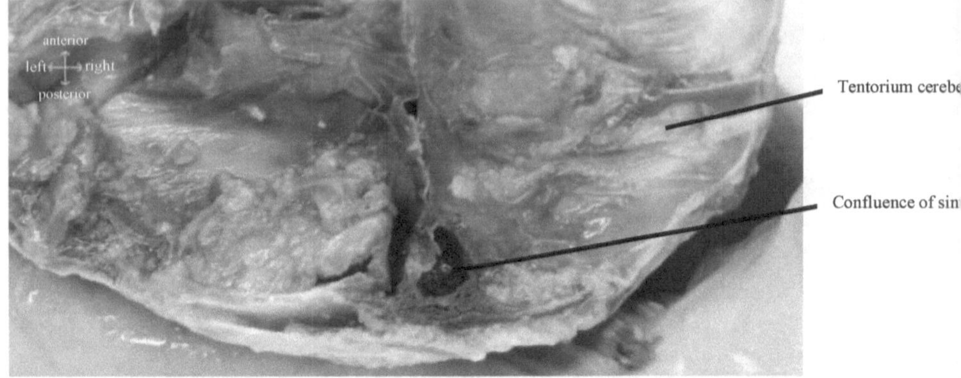

Figure 3.4: Base of the skull showing the exposed confluence of sinuses. Image from the current research project.

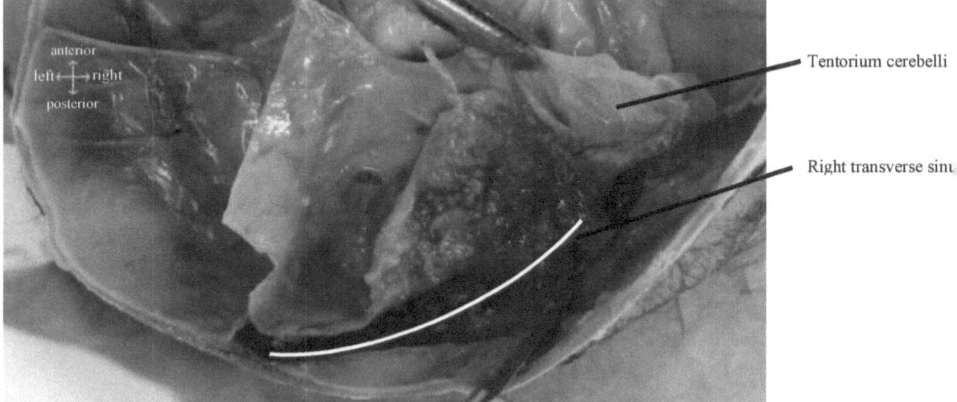

Figure 3.5: Base of skull with the tentorium cerebelli opened to expose the right transverse sinus (white line). Image from the current research project.

3.2.2 Children, adolescents and adults

Contrast enhanced magnetic resonance images from Red Cross Children's War Memorial (infants and children) and Groote Schuur Hospitals (adolescents and adults) were grouped based on age and sex. The term sex is used to refer to the biological sex (male or female) of an individual as documented in the patient folder.

Age groupings are based on current South African legislation. According to the Children's Act, 2005 (Act 38 of 2005), children are defined as any person younger than 18 years of age. Persons 65 years and over are referred to as older persons in terms of the Older Persons Act, 2006 (Act

13 of 2006). Persons between the ages of 18 and 65 are referred to as adults for the purposes of this study; as shown below, group A consisted of children and groups B - E of adults.

Five groups were created in order to enable a credible comparison between the various ages:

Group A - Persons 0 – 18 years (infants, children and adolescents)

Group B - Persons 19 to 25 years (young adulthood)

Group C - Persons 26 to 55 years (before midlife)

Group D - Persons 56 to 64 years (after midlife)

Group E - Persons 65 years and older (elderly people)

3.2.3 Variation of the dural venous sinuses

Based on the available literature and on our current observations of the radiological images variations in the sample were grouped as follows, with the term multiple being used to refer to more than one complete structure ipsilaterally:

1) Absence of the transverse sinus on the right
2) Absence of the transverse sinus on the left
3) Absence of the transverse sinuses bilaterally
4) Multiple transverse sinuses unilaterally
5) Multiple transverse sinuses bilaterally
6) Right dominant transverse sinus
7) Left dominant transverse sinus
8) Codominance of the transverse sinuses
9) Multiple superior sagittal sinuses
10) Bifurcation of the SSS into left and right transverse sinuses (no confluence)
11) Bifurcation of the SS into left and right transverse sinuses
12) SSS draining into only one of the transverse sinuses
13) SSS draining into only one of the transverse sinuses and the SS draining into the other
14) Absence of sigmoid sinus on the right
15) Absence of the sigmoid sinus on the left
16) Absence of sigmoid sinus bilaterally
17) Absent inferior sagittal sinus
18) Absent straight sinus
19) Multiple straight sinuses
20) Variations in the configuration of the confluence of sinuses, including partitioning into compartments

Refer to section 4.3.7, 4.71 and 4.9 for complete results on which of the above variations were found in the current study.

3.2.4 Variations at the confluence of sinuses

The confluence of sinuses was visually inspected, and variations grouped as follows (based on the study by Gökçe *et al.,* 2014); refer to table 3.1 below.

Variation types I and III were kept the same, however, type II was too limiting for the study and was expanded to include all variations in the configuration of the confluence of sinuses.

Table 3. 1 Table comparing the current study to the study done by Gökçe *et al.* in 2014.

	Current study	**Study by Gökçe *et al.,* 2014**
Type I	True confluence (SSS + SS + RT+ LT)	True confluence (SSS + SS + RT+ LT)
Type II	A partial confluence – bifurcation of the SSS and/or the SS into the transverse sinuses, thus forming a ring structure in the confluence of sinuses	Only three of the sinuses named in type 1 drained into the confluence of sinuses
Type III	Non-confluence	Non-confluence

Bayaroğullari *et al.,* (2018); further divided the above types into subtypes, listed below. For the current study the type I, II and III as described above was found to be sufficient.

Classifications of anatomical variations Type I
Type IA (SSS + SS + R-TS + L-TS)
Type IB (SSS + SS + R-TS + Hypo.* L-TS) ± OS
Type IC (SSS + SS + Hypo. * R-TS + L-TS) ± OS
Type ID (SSS + SS + Bil. Hypo. ** TS) ±OS

Classifications of anatomical variations Type II
Type IIA1 (SSS + SS + R-TS/SSS + L-TS)
Type IIA2 (SSS + R-TS/SSS + SS + L-TS)
Type IIB1 (SSS + SS + R-TS/Ag.*** L-TS) ± OS
Type IIB2 (SSS + SS + L-TS/Ag.*** R-TS) ± OS
Type IIC (SSS + SS + R-TS + L-TS)
Type IID1 (SSS + SS + R-TS/SS + Hypo.* L-TS) ± OS
Type IID2 (SSS + SS + L-TS/SS + Hypo.* R-TS) ± OS
Type IIE1 (SSS + SS + R-TS/SSS + L-TS
Type IIE2 (SSS + SS + L-TS/SSS + R-TS)

Classifications of anatomical variations Type III
Type IIIA (SSS + SS + L-TS/SSS + R-TS) ± OS
Type IIIB (SSS + L-TS/SS + R-TS) ± OS

* Hypoplastic ** Bilaterally hypoplastic *** Agentic (sic) or absent

3.2.5 Determining dominance

Two main approaches for determining dominance of cerebral venous structures were found in the literature (refer to section 2.8 above).

The first approach was described by Widjaja & Griffiths (2004), and also used in the study by Surendrababu & Livingstone, 2006. The transverse sinuses were measured 1 cm from the confluence of sinuses on either side and the SSS was also measured 1 cm from the confluence.

The second approach is by Kitamura *et al.,* (2017). This author defined the midline as a line perpendicular to the mean distance between the medial border of the orbital rims. The transverse sinus measurement was taken between the ninth and tenth portions of the distance between the edges of each side; while sigmoid sinus was measured at the average distance of the length of the sinus. Kitamura *et al.,* used measurement ratios between left and right to determine dominance. If the ratio was more than 1.5 the sinus was classified as dominant right, if the ratio was less than 0.67 if was considered dominant left (thus, sinus was more than 50% that of the contralateral side). However, if the ratio fell between 1.5 and 0.67 the sinuses were considered to be symmetrical (of equal dominance).

As time-of-flight MRVs could not be obtained for all patients within the sample of this study, the internal jugular veins and jugular bulb height could not be measured or included in the statistical analyses.

3.2.5.1 Current approach

Volume tracing has been done for different structures in the brain, however studies conducting tracing on venous structures could not be found in the English literature.

For the current study it was decided to measure the volume of the following sinuses:

- Right and left transverse sinuses
- Right and left sigmoid sinuses
- Superior sagittal sinus

More robust data for statistical analysis and comparisons in structure may be obtained by using the volumes of the sinuses rather than by measuring only one or two diameters of sinuses. The literature (Luigi *et. al.,* 2007; Egemen & Solaroglu, 2017) also shows that the transverse and SSS tend to be more triangular in shape and not circular; thus, depending on the level of the measurement one might not get a clear view of the width of the sinus. By using volume tracing, one may evaluate the entire sinus in size and shape, to gain a better understanding of blood flow through the sinus.

Dominance was determined by using a modified approach to that of Kitamura *et al.,* (2017). The current study used the entire volume of the transverse sinus and not just a diameter measurement when determining cerebral dominance of an individual. Dominance was expressed in a ratio of the right to left transverse sinuses, if the ratio was more than 1.5 the sinus was classified as dominant right, if the ratio was less than 0.67 if was considered dominant left (thus, sinus was more than 50% that of the contralateral side). However, if the ratio fell between 1.5 and 0.67 the sinuses were considered to be symmetrical (of equal dominance).

As the study uses the volume of the transverse sinuses to calculate dominance, one should consider the effect that accessory vessels draining into these sinuses might have on determining dominance. The researcher hypothesised that the drainage of accessory vessels into the sinuses will increase the volume of the respective sinuses, however as the volume of the transverse sinus already includes the blood flow from the accessory vessels, it will not influence the cerebral dominance.

3.2.6 Hypoplasia and aplasia

According to the study done by Widjaja and Griffiths (2004) the transverse sinus was said to be hypoplastic when the diameter of the transverse sinus was less than half the diameter of the SSS. Widjaja and Griffiths (2004) measured the transverse and SSS 1 cm from the confluence of sinuses.

The method used by these authors was adjusted and extrapolated for volume measurements and used in the current study. Thus, if the volume of the transverse sinus was less than half the volume of the SSS the transverse sinus was classified as being hypoplastic.

The current study only used post contrast imaging for analysis, thus if a structure could not be visualised at all it was classified as aplastic, as the presence of contrast in the vessel would enable the viewer to see the smallest of vessels. Vessels were also examined in three different planes (coronal, sagittal and transverse), to ensure the most holistic view of a vessel is documented.

3.2.7 Venous blood pools

In the current study the following areas were carefully inspected for any signs of dural venous blood pools:

- the caudal end of the roof of the fourth ventricle
- area around the brain stem and pons (the junction of the midbrain and hindbrain)
- area around the thalamus and hypothalamus (diencephalon along the caudal margin of the cerebral hemisphere)
- within the loose tissue composing the tentorium cerebelli

3.2.8 Tracing protocol

3.2.8.1 Software

The free software Horos (The Horos Project, licensed under the GNU Lesser General Public License, Version 3.0), was used to trace the outline of the dural venous sinuses and construct a 3D image of each venous sinus.

The software allowed the researcher to apply a red and yellow colour CLUT (colour look-up table) filter to the MRIs. The filter provided better contrast between the vessels and the surrounding brain tissue. The filter did not distort or change the original MRI in anyway, it is purely a preference of the researcher to better view the blood vessels and do the venous tracings.

3.2.8.2 Anatomical landmarks and measurements of the sinuses

General considerations

As sinuses are variable from person to person, the approach had to be adapted to capture an accurate representation of the dural sinuses of each person; while applying a consistent methodology. In order to ensure a uniform approach to the volume measuring the following general guidelines were taken into account.

- After volume calculations the 3D structure was visually inspected to confirm that it conforms to the generally accepted structures described in *Gray's Anatomy* and the literature in general. If a discrepancy was found, the volume tracing was re-done. In all cases where this was done the discrepancy was no longer present and the volume measurement was included in the data.
- The measurements for the transverse sinuses were started in the midline of the cerebrum (where a patent aqueduct of the midbrain is visualised). Thus, the first measurement for the left and right transverse sinuses are the same.
- In cases where there was any doubt as to the exact diameter of the sinus, a more conservative approach was taken, whereby the clearly visible internal diameter of the sinus was used so as not to artificially increase the volume of the sinus, refer to figure 3.6 and 3.7 below.
- Sinuses are not uniform in cross-sectional shape throughout their length, thus careful consideration as to the anatomy of the structure, the plane in which one is viewing the sinus and the measurements before and after each point of measurement should always be taken into consideration to ensure that the most accurate representation of the sinus is being made.
- Volumes of the sinuses were recorded up to four decimal places, as this is the format in which the software used for tracing generated the values.

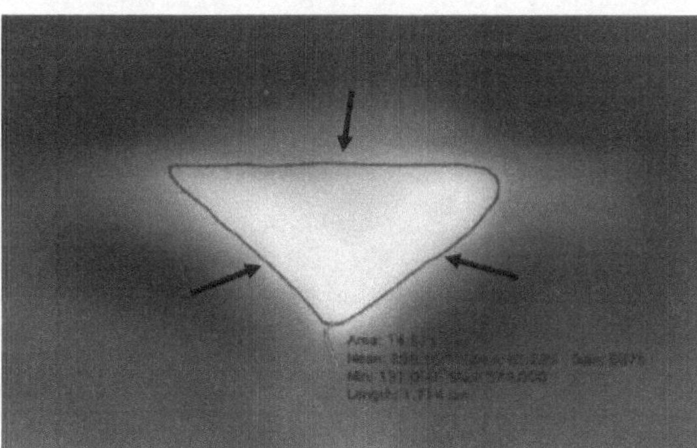

Figure 3.6: Enlarged view of a conservative tracing of the outline of the SSS. The tracing line is placed at the edge of the brightest yellow part, which represents the contrasted blood flow through the sinus. The black arrows are indicating the fading border of the SSS, as it cannot be exactly determined where the wall of the SSS is. Image from the current research project.

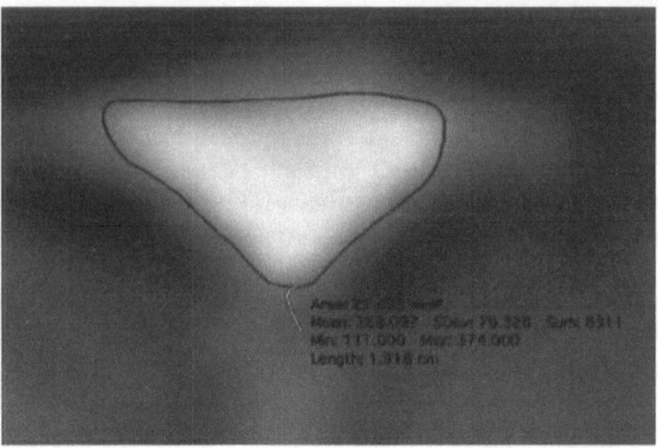

Figure 3.7: Enlarged view of a non-conservative tracing of the outline of the SSS. The tracing line is placed on the outside of the yellow, enclosing the lumen and wall of the SSS. No yellow remnants are to be seen outside of the line. Using this tracing method may overestimate the volume of the sinus as it includes the wall and might include some of the surrounding tissue within the line. Image from the current research project.

Transverse sinuses

The left and right transverse sinuses start at the confluence of sinuses and continues laterally towards the posterolateral part of the petrous part of the temporal bone, where they turn inferiorly to become the sigmoid sinuses. The sinus runs in the attached margin of the tentorium cerebelli, first on the squamous part of the occipital bone and then on the mastoid angle of the parietal bone. The transverse sinus follows a gentle antero-lateral curve, increasing in size as it flows towards the sigmoid sinus. Typically, the right transverse sinus is larger than the left sinus and drains blood from the superficial parts of the brain; while the left drains blood mainly from the deep parts of the brain. Usually, the pattern of drainage of the right sinus is continuous with the SSS and the left sinus with the SS. For the current study the transverse sinus was measured from the midline of the confluence (line A) to the posterolateral part of the petrous part of the temporal bone (line B), see figures 3.8 – 3.11 below.

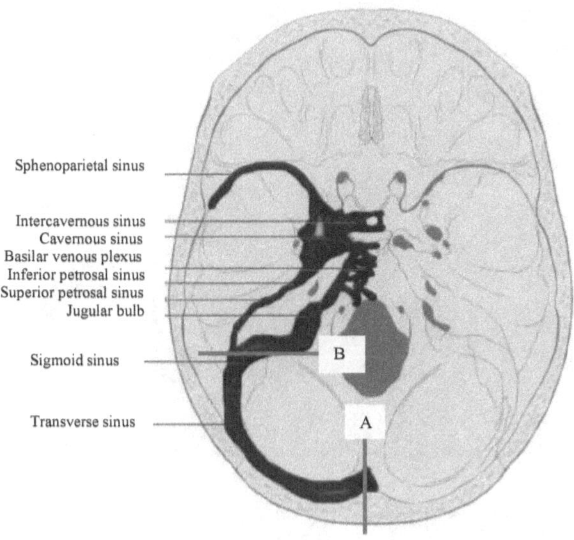

Figure 3.8: Measurement of the transverse sinus, with **A** – midline of the confluence and **B** – Posterolateral part of the petrous part of the temporal bone. Adapted from the *Dural Venous Sinuses,* in *Radiopaedia,* courtesy of A. Prof. Frank Gaillard. Retrieved December 26, 2019 from .Used with permission.

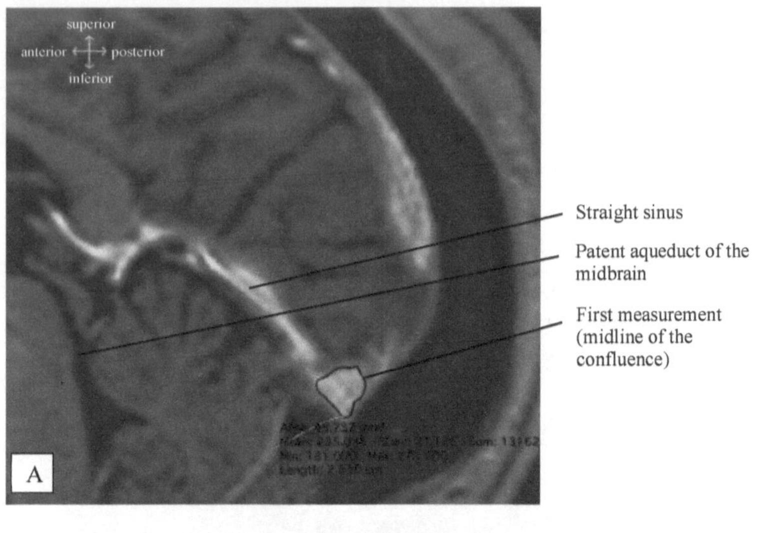

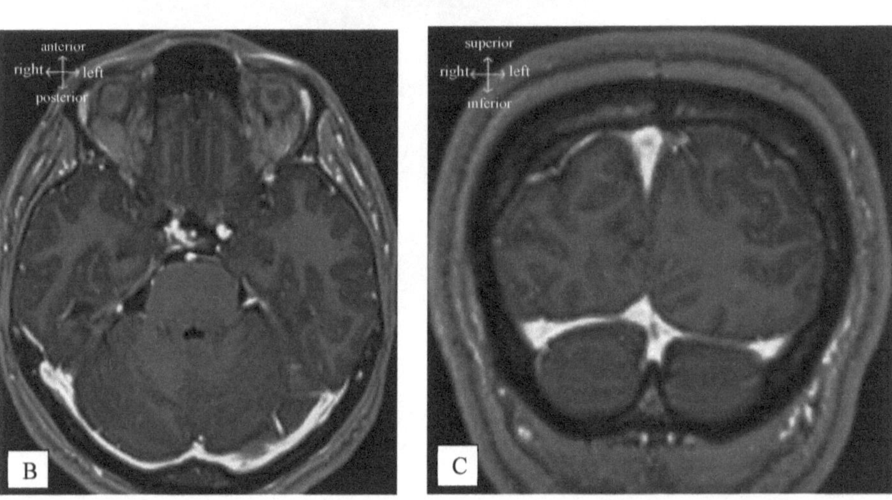

Figure 3.9: The first measurement of the left transverse sinus as seen in the sagittal view (image A) on a contrast enhanced MR image. **B and C:** Horizontal and coronal planes were used to orientate the researcher as to their position within the sequence of MR slices. The slices seen in B and C correspond to the first measurement on image A. Image from the current research project.

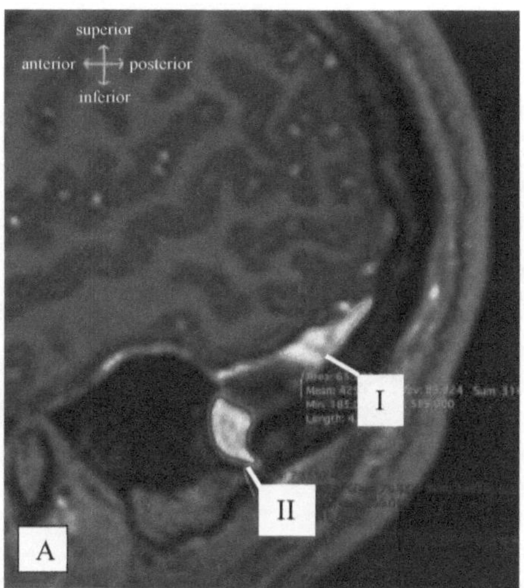

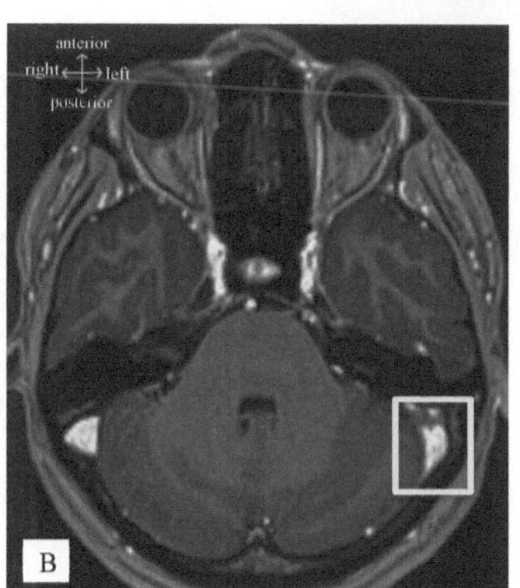

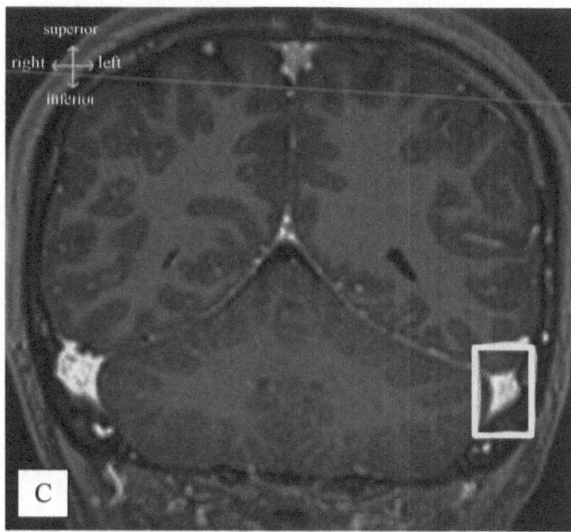

Figure 3.10: The last measurement of the left transverse sinus as seen on a contrast enhanced MR image. **A:** Position of last measurement in the sagittal view (marked I on figure A). The dark outlined measurement, marked II, is included with the tracing of the sigmoid sinus. **B and C:** Horizontal and coronal (last measurement of the left transverse sinus outlined in yellow) planes were used to orientate the researcher as to their position within the sequence of MR images. The images seen in B and C correspond with the last measurement on slice A. Images from the current research project.

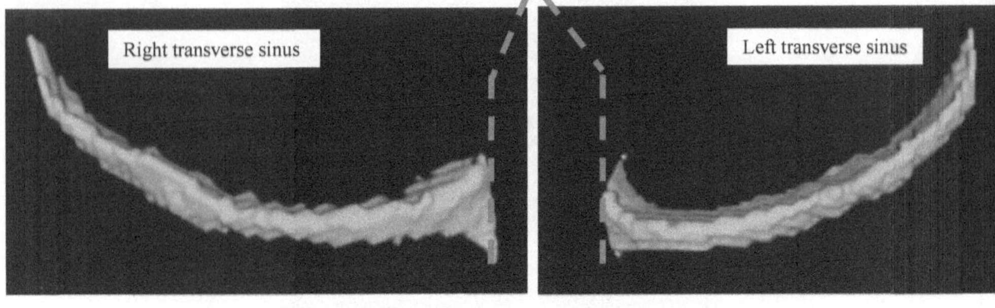

Figure 3.11: Three-dimensional structure of the right and left transverse sinuses. Note, images on MRIs are always the inverse of macroscopic structure; thus, the RT "appears" to go to the left while the LT "appears" to go to the right. Images from the current research project.

Sigmoid sinuses

The transverse sinuses continue as the sigmoid sinuses, beginning where the transverse sinus leaves the tentorium cerebelli on either side of the cerebrum. Curving inferomedially in a groove on the mastoid process of the temporal bone, each sigmoid sinus crosses the jugular process of the occipital bone and turns anteriorly to the superior jugular bulb, lying posterior to the jugular foramen. The sigmoid sinuses may be found to be larger than the transverse sinuses due to the draining of the inferior and superior petrosal sinuses into the sigmoid sinuses.

In the current study measurements of the sigmoid sinuses were taken from where the transverse sinus leaves the tentorium cerebelli until the sigmoid sinus reaches the superior jugular bulb, see figures 3.12 to 3.14 below.

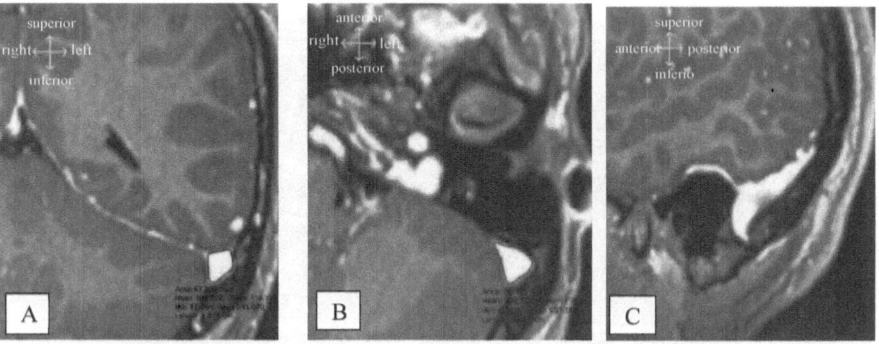

Figure 3.12: First measurement of the sigmoid sinuses. **A** – Coronal plane, **B** – Transverse plane, **C** – Sagittal plane. Images from the current research project.

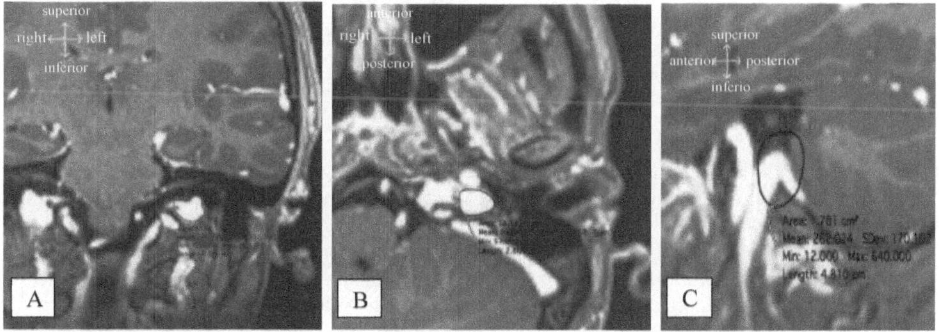

Figure 3.13: Last measurement of the sigmoid sinuses. **A** – Coronal plane, **B** – Transverse plane, **C** – Sagittal plane, indicating the reference point in the sagittal plane used to determine when to stop measuring the sigmoid sinus. Images from the current research project.

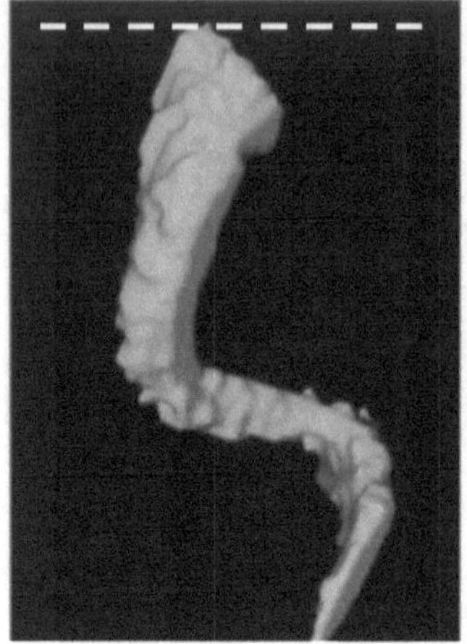 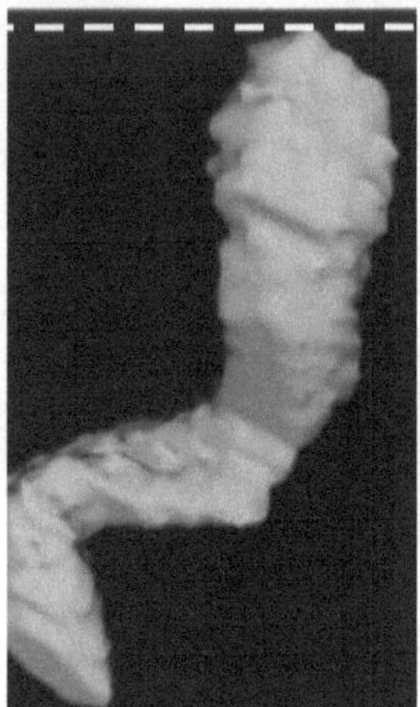

Figure 3.14: Three-dimensional structure of the right and left sigmoid sinuses. Images from the current research project.

Superior sagittal sinus

Running in the convex margin of the falx cerebri, the SSS grooves the internal surface of the frontal bone, the adjacent margins of the two parietal bones and the squamous part of the occipital bone. The sinus begins near the crista galli just a few millimetres posterior to the foramen caecum. The cross section of the sinus is characteristically triangular in shape, with the apex directed inferiorly and continuous with the falx cerebri. The SSS starts narrow anteriorly and gradually widens as it runs posteriorly and finally enters the confluence of sinuses.

The current study measured the SSS from its origin near the crista galli anteriorly towards the confluence of sinuses. The first tracing was made as soon as the SSS was visible in the coronal plane, looking from anterior to posterior, see figure 3.15 to 3.18 below.

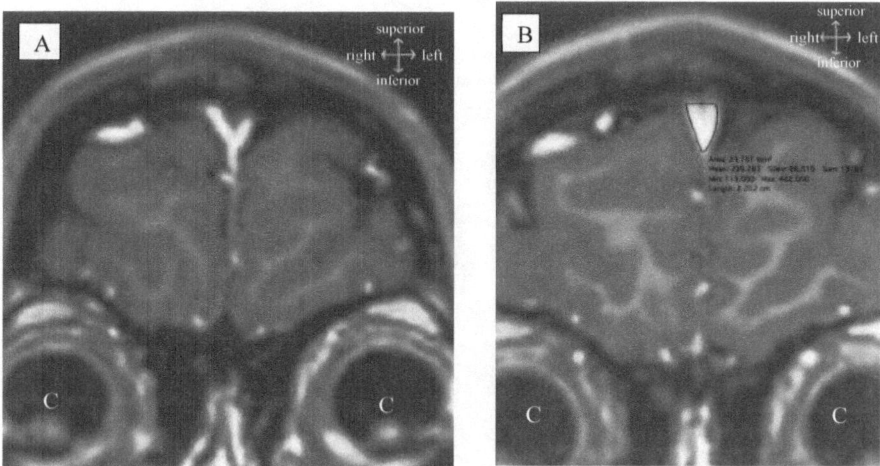

Figure 3.15: First measurement of the SSS. Image A (coronal plane) shows the MR slice just before the triangular shape of the SSS is visible. Image B (coronal plane) shows the first volume tracing made for the SSS. The bony orbits are also visible (point **C** on both images). Images from the current research project.

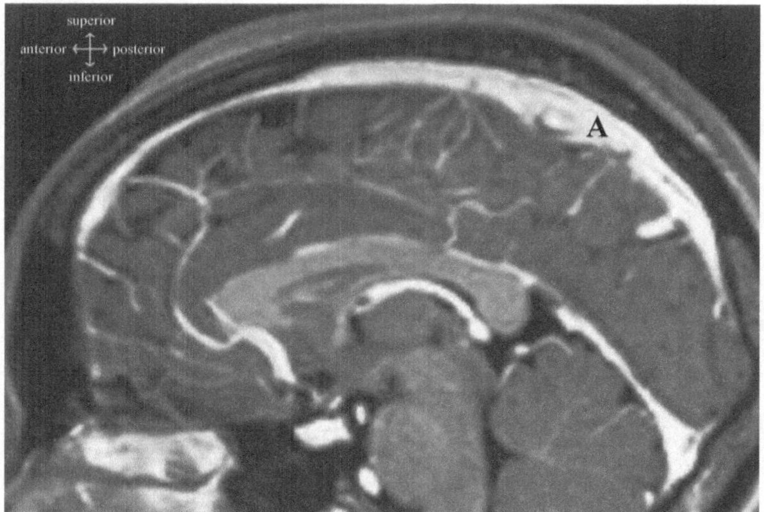

Figure 3.16: The SSS (**A**) as seen on a contrast enhanced MR image in the sagittal plane. Image from the current research project.

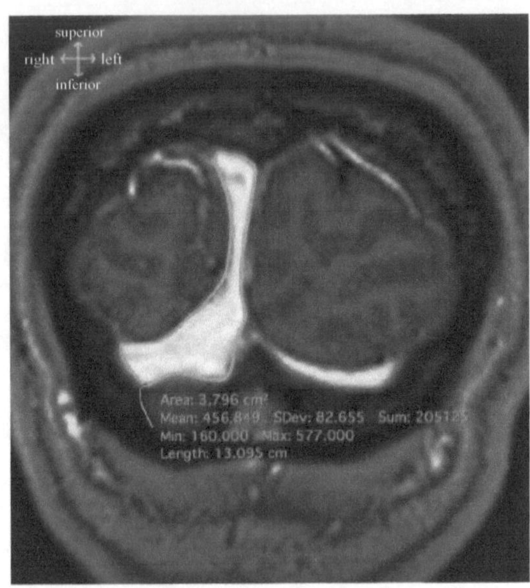

Figure 3.17: The last measurement of the SSS (green tracing) as seen on a contrast enhanced MR image in the coronal plane. Image from the current research project.

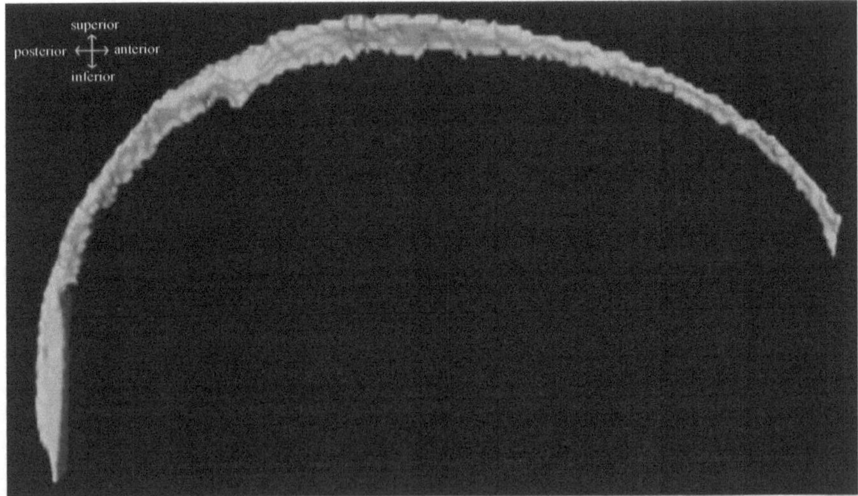

Figure 3.18: Three-dimensional structure of the SSS showing the total volume of the sinus. Image from the current research project.

Cerebral veins

To visualise the superficial veins of the brain, a maximum intensity projection (MIP) filter was applied to the image within the imaging software (figure 3.18 &3.20). The MIP is used to enhance the appearance of the blood flow through the veins and thus better visualizing the vessel. As we are only determining presence of the vessel any influence that the MIP might have on the circumference and thus the volume of the vessel is negligible, see figure 3.19 – 3.22 below.

<u>Veins of Trolard and Labbé</u>

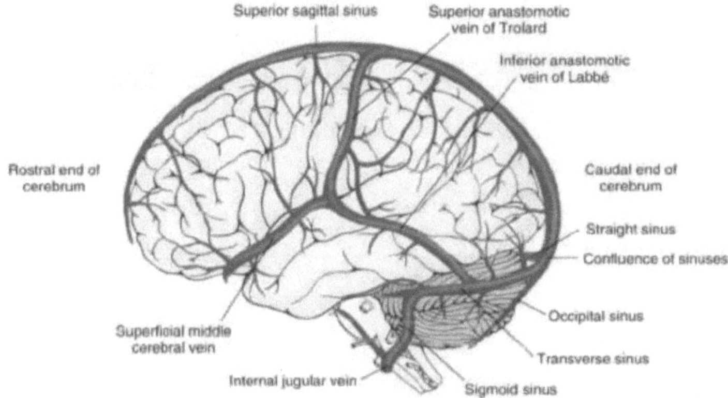

Figure 3.19: Illustration of the superficial veins of the brain. Retrieved from on 06 June 2020. Used with permission.

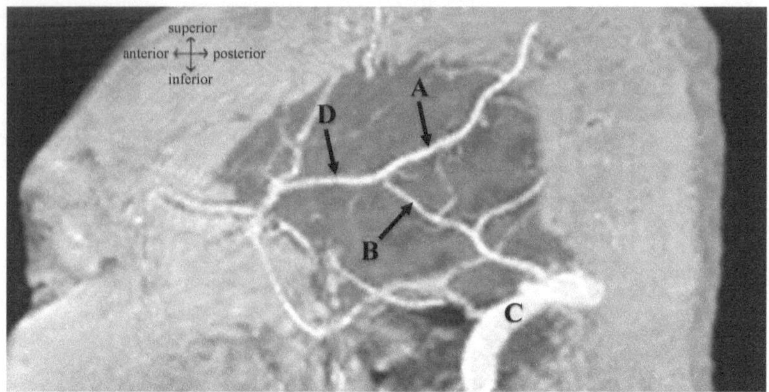

Figure 3.20: Image of the vein of Trolard (**A**) and the vein of Labbé (**B**). The transverse sinus (**C**) and the superficial middle cerebral vein (**D**) can also be seen. Image in the sagittal view from the current research project.

<u>Vein of Galen and the internal cerebral veins</u>

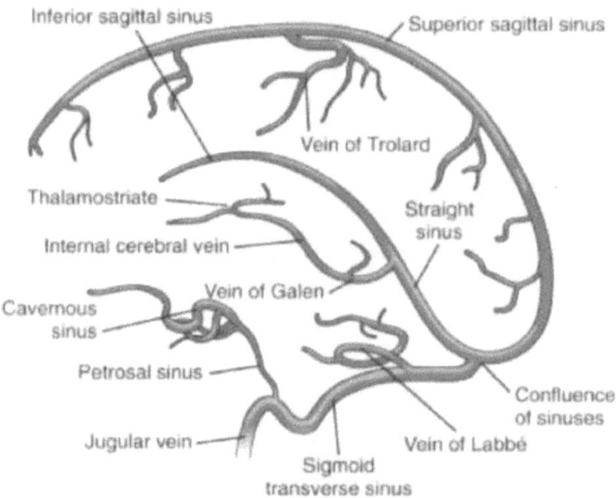

Figure 3.21: Schematic drawing showing the vein of Galen and the internal cerebral veins. From *Pharmacology and Physiology for Anesthesia,* (page 174) by B.P. Lemkuil *et al.,* 2019, Philadelphia, PA: Elsevier. Used with permission.

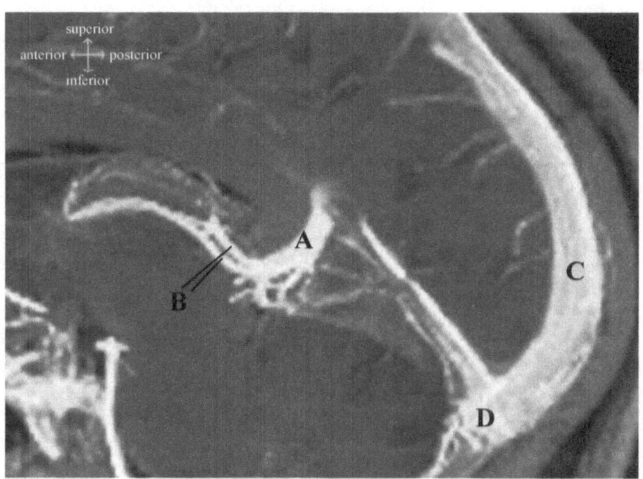

Figure 3.22: Image showing the vein of Galen (**A**) and the paired internal cerebral veins (**B**). The SSS (**C**) and confluence of sinuses (**D**) are also visible. Image in the sagittal view from the current research project.

3.2.9 Image analyses

All images were assessed for normal anatomy and anatomical variations. The free, online, software program Horos (The Horos Project, licensed under the GNU Lesser 3D model was created for each participant, refer to figure 3.2 below. The 3D model was then reviewed, and variations and/or embryonic structures described.

Sinuses were reviewed in three planes (coronal, transverse and sagittal), due to the literature indicating that some sinuses may not be visualised completely when only looking in one plane (Rollins *et al.,* 2005).

Images were evaluated for anatomical variations of dural venous sinuses and presence of accessory sinuses.

The dural venous sinuses included in this study are the SSS, sigmoid sinuses, transverse sinuses and SS. The occipital sinus was excluded as its borders could not accurately be determined on the magnetic resonance images. A summary of the investigations done for each sinus is listed in table 3.2 below.

Table 3. 2 Table showing variations that were noted for the current study.

Dural venous sinus	Variations
Superior sagittal sinus	Presence and/or absence of the sinus Bifurcation of the sinus
Transverse sinuses	Left, right or equal dominance Hypoplasia and aplasia
Inferior sagittal sinus	Unilateral and bilateral presence and/or absence of the sinuses Hypoplasia and aplasia
Sigmoid sinuses	Unilateral and bilateral presence and/or absence of the sinuses Hypoplasia and aplasia

3.2.10 Venous sinus configuration as a method of identification

Building on the work of R.S. Tubbs, 2019, in the current study a method was proposed to apply venous sinus configuration and drainage patterns from the perspective of forensic clinical anatomy.

Each variable was assigned a number, therefore generating a sequence of numbers which were referred to as the identification number (I-Number).

Table 3. 3 Table used for assigning of an identification number (I-Number).

Sex	Male	1
	Female	2
Hypoplasia	None	0
	Both sides	1
	Right side	2
	Left side	3
Dominance	Right side	1
	Left side	2
	Co-dominant	3
SSS drainage pattern (continued overleaf)	Generally acceptable anatomical configuration	1
	The SSS drains only into the RT, with the SS draining into the RT and LT	2
	The SSS drains only into the LT, with the SS draining into the RT and LT	3

	The SSS bifurcates and drains into the RT and LT respectively and not into the confluence of sinuses.	4
SSS drainage pattern (continued)	The SSS drains into the right transverse sinuses and the SS drains into the left transverse sinus (SSS to the RT and the SS to the LT)	5
	The SSS drains into the left transverse sinuses and the SS drains into the right transverse sinus (SSS to the LT and the SS to the RT)	6
	The SSS drains to both the RT and LT while the SS drains only to the LT	7
	The SSS drains to both the RT and LT while the SS drains only to the RT	8
	The SSS and SS drain to the same side	9
Variation of the confluence of sinuses	True confluence	1
	Partial confluence	2
	Non-confluence	3
Other variations	No other variation	0
	Absent sigmoid sinus	1
	Absent straight sinus	2
	Absent superior sagittal sinus	3
	Absent transverse sinus	4
	Absent inferior sagittal sinus	5
	Multiple sigmoid sinuses	6
	Multiple straight sinuses	7
	Multiple superior sagittal sinuses	8
	Multiple transverse sinuses	9
	Multiple inferior sagittal sinuses	10

Thus, if a patient was a female with no hypoplasia, co-dominant, a type 4 SSS drainage pattern, with a non-confluence and no other variations, the following I-Number would be assigned: **203430**

3.3 Statistical Analyses

Data were organised in spreadsheets using Microsoft Excel (v. 16.13, Microsoft Corp, 2018). Statistical analyses were done using Stata Ver. 13 (StatCorp, TX, USA). The statistical analyses were completed with assistance from statistical consultant at the University of Cape Town. He has more than six years' experience in lecturing forensic statistics at the University of Cape

Town, along with being the in-house statistical consultant for MPhil and MMed research studies.

The variables could be grouped as epidemiological (age and sex) or as angiographic (volume of sinus, presence/absence of sinus and division of SSS).

3.3.1 Intra-observer error

In order to minimise the intra-observer error, the following was done:

1) Ten percent of the sample (32 images) were initially measured and were then re-measured by the researcher.
2) The measurements were analysed for a statistically significant difference between the two sets of measurements.
3) No statistically significant difference was found between the two sets of measurements done by the researcher, refer to table 3.4 below. Therefore, measurements could be taken on the remaining 90% of the images, with confidence regarding their accuracy.

Table 3. 4 Table indication the technical error of measurement (TEM) and the coefficient of reliability (Cohen's Kappa coefficient was used) used to determine the intra-observer error.

Vessel	TEM	coefficient reliability
Superior sagittal sinus	0.2573	0.9508
Right Transverse sinus	0.1835	0.9398
Left Transverse sinus	0.1389	0.9374
Right Sigmoid sinus	0.2921	0.9089
Left Sigmoid sinus	0.3025	0.7789

Ideally the coefficient of reliability should be as close to one and technical error of measurement TEM as close to zero as possible. The closer the values to one and zero respectively, the higher is the confidence that the measurements were accurately taken.

3.3.2 Inter-observer error

In order to minimise the inter-observer error, the following was done:

1) A person external to the study was trained to take the necessary measurements
2) A person external to the study directed the placing of the cursor on the images by the researcher and subsequent recording of results of five random images (3,5 hours of tracing)
3) The measurements by the person external to the study were compared to those of the researcher and found to be within one or two decimal places of accuracy
4) Therefore, it was shown that if a second person were to use the current protocol for taking these measurements the outcomes would be virtually identical

3.3.3 Statistical tests performed

3.3.3.1 Analysis overview

Shapiro Wilk tests were used to determine if numerical variables were normally distributed. Neither age nor any of the volumes of the vessels were normally distributed and therefore non-parametric tests were conducted.

Differences within numerical variables were assessed using the Wilcoxon Rank Sum test (for two group comparisons) and the Kruskal Wallis test (for multi-group comparisons). Where necessary pairwise post-hoc testing was conducted with Bonferonni correction applied.

3.3.3.1 Descriptive statistics

Descriptive statistics were done on the sample, this included the following:
- Distribution by age
- Distribution by sex
- Distribution by presence of variation
- Distribution by presence of pathology

3.3.3.2 Chi-square test

Categorical variables were analysed using the Pearson's Chi-squared test.

Chi-Square test were done to review the following:
- Distribution of variation by sex
- Distribution of variation by age group
- Distribution of pathology by sex
- Distribution of pathology by age group
- Presence of variation and presence of pathology

3.3.3.3 Venous sinus volume

The venous sinus volume obtained by tracing the dural venous sinus was used to analyse the following:
- Venous sinus volume and age
- Venous sinus volume and sex
- Venous sinus volume and presence of pathology
- Venous sinus volume and presence of accessory vessels
- Venous sinus volume and variation in the drainage pattern of the SSS
- Venous sinus volume and variation of the confluence of sinuses
- Venous sinus volume and hypoplasia
- Comparison between venous sinus volume of the sinuses on the right side and venous sinuses on the left side

In general in the animal kingdom females are larger than males. However, the opposite is true for mammals and birds, where males are most often larger than females (Gustafsson & Lindenfors, 2004). A size difference may also be seen when comparing the human brain between males and females, with males having a 10% heavier brain than females (Dekaban,

1978). Vessel volumes for males were therefore reduced by 10% and all relevant data re-analysed to determine if the difference in size between males and females may be the underlying reason for the differences seen between subjects when analysed by sex.

Chapter 4: Results

4.1 Fetal dissection

Stillborn fetuses of full-term age were obtained from the New Somerset Hospital. In total 5 fetuses were dissected. The dissection of the fetuses was to confirm what has been reported in the literature and to gain a better understanding of the venous structure in new-borns, and they were thus not included in the study.

The venous sinuses were reviewed in situ. No variations were noted in the five fetal dissections. Figure 4.1 below shows the transverse sinuses covered by the tentorium cerebelli and the point at which the SSS drains into the confluence of sinuses.

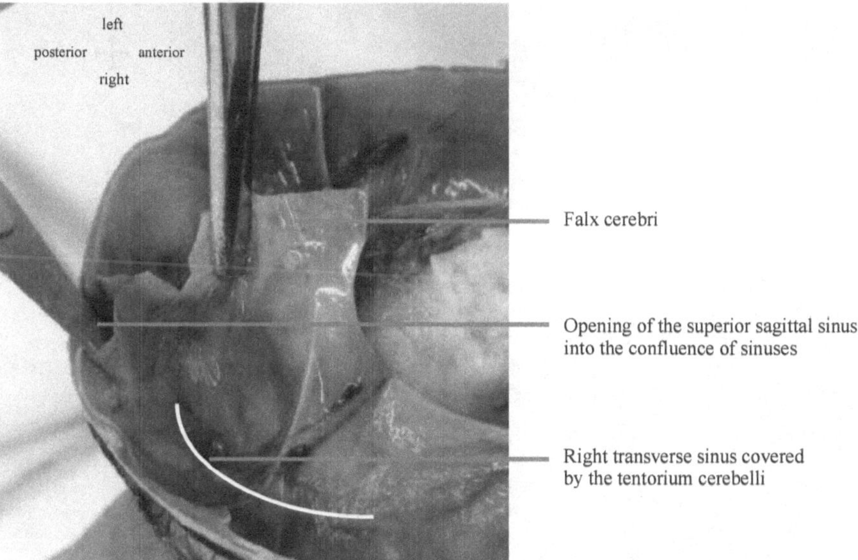

Figure 4.1: Opening of the confluence of sinuses by forceps. The transverse sinuses are covered by the tentorium cerebelli. Image from the current research project.

The right (figure 4.2) and left (figure 4.3) transverse sinuses were opened by cutting through the tentorium cerebelli into the sinus below. No variation was seen within the drainage pattern of the transverse sinuses from the confluence of sinuses.

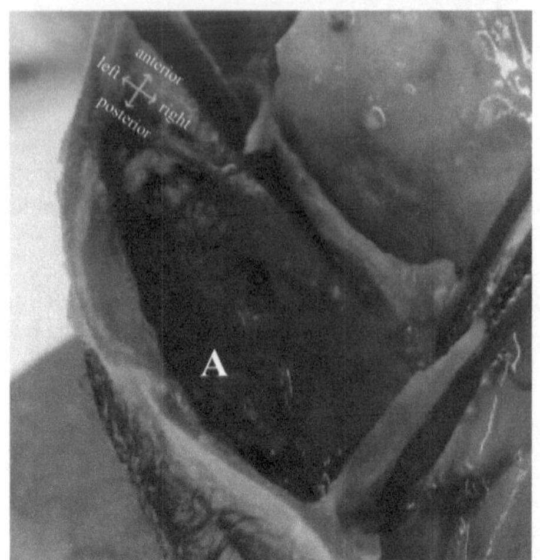

Figure 4.2: Dissection of the right transverse sinus (**A**). The tentorium cerebelli was resected to expose the sinus below. Image from the current research project.

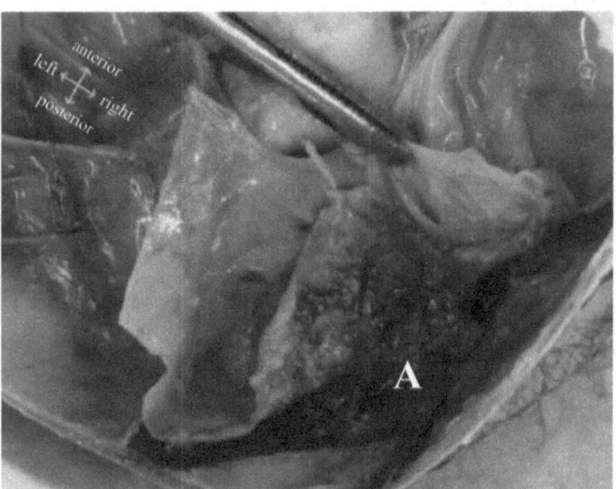

Figure 4.3: Dissection of the left transverse sinus (**A**). Image from the current research project.

The SSS could be seen draining into the confluence of sinuses. None of the fetuses had any variation of the SSS. The superior sagittal and transverse sinuses can still be seen covered in dura on figure 4.4 below.

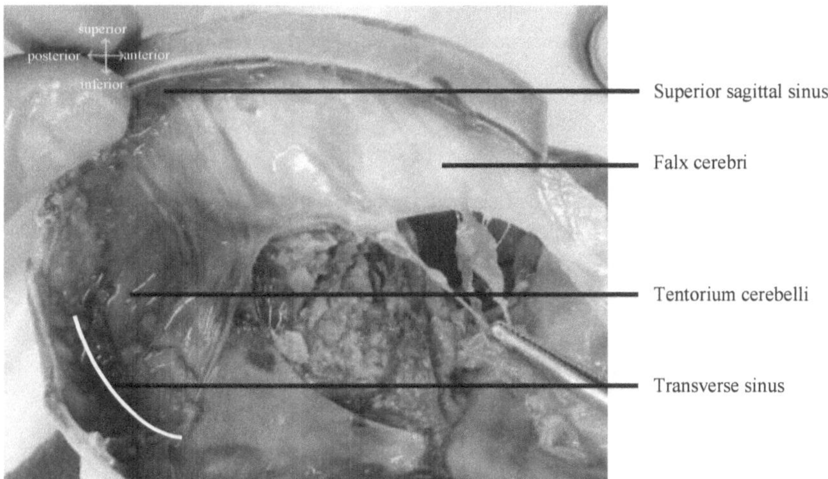

Figure 4.4: The superior sagittal and transverse sinuses enclosed in dura mater, draining into the confluence of sinuses posteriorly. Image from the current research project.

The SSS was seen to be contained within the dura mater. Figure 4.5 below shows the SSS still enclosed in dura mater.

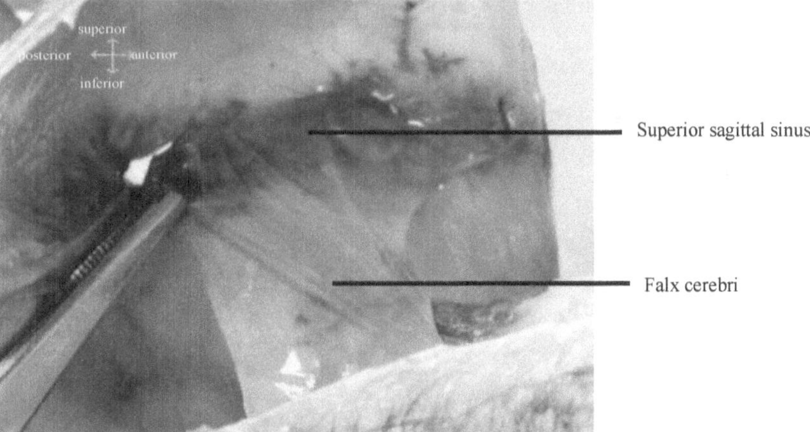

Figure 4.5: The SSS enclosed in dura mater. Image from the current research project.

<u>**4.2 Epidemiological variables**</u>

For the current study the following definitions applied:

Mean – Refers to the sum of terms divided by the number of terms, also referred to as the average. The mean may be more susceptible to outliers and is generally used when the data is normally distributed (Lund & Lund, 2020).
Median – Refers to the middle value in a numerical list. The median is not as susceptible to outliers as the mean and is generally used when the data is not normally distributed (Lund & Lund, 2020).

As most data in the current study was found to be not normally distributed the median was used in the analyses, however, where applicable the mean value may also be listed to give a complete view of the data set.

4.2.1 Sex

Sex was distributed as follows: 121 males and 181 females (refer to graph 4.1 below).

4.2.2 Age

The mean age of all subjects was 45.39 years (standard deviation of 15.96). The minimum age was 13 years and the maximum age was 81 years. The median age of females (50 years) in the study is significantly greater than the median age of males (43 years) (p= 0.0143).

Age was categorised in two ways. First, in terms of life stage as described in section 3.2.2 (Graph 4.2), where it was noted that the overwhelming majority of persons were between 26 and 55 years of age (before midlife). Second, age was categorised in ten-year intervals (Graph 4.3) in order to have even distribution of subjects across all of the sub-groups in preparation for the data analyses.

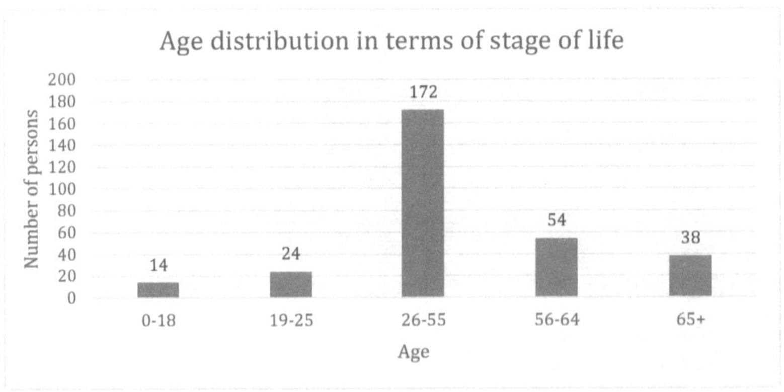

Graph 4. 1 Age distribution in terms of stage of life.

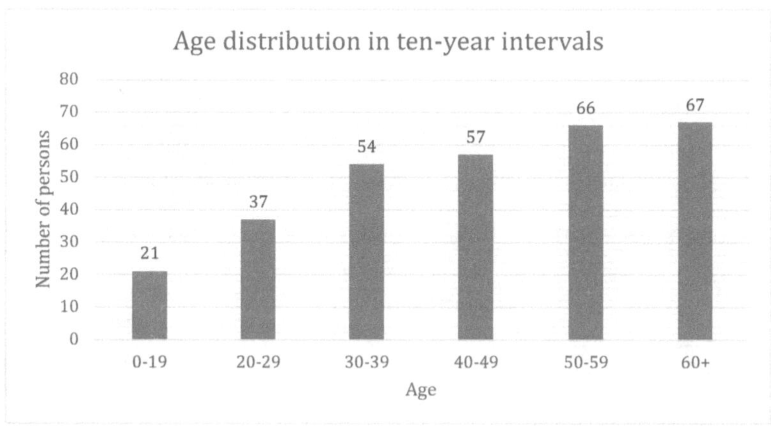

Graph 4. 2 Age distribution in ten-year intervals.

<u>**4.3 Angiographic variables**</u>

4.3.1 Venous sinus volume

The data for the volumes of the venous sinuses were found to be non-normally distributed by applying a Shapiro Wilks test for normality. Thus, non-parametric tests were applied for data analysis.

The superior sagital sinus was found to have the greatest volume (10.8275 cm^3), followed by the right sigmoid (6.8563 cm^3) and left transverse sinuses (6.5815 cm^3) .

Table 4. 1 Descriptive statistics showing the minimum, maximum and mean volume in of the cerebral venous sinuses (copied from statistical analyses report for the current study).

```
    stats |      volrt      vollt    volsigr    volsigl     volsss
----------+----------------------------------------------------------
        N |        300        297        299        295        301
     mean |     1.9187   1.553675   2.687484     2.1451   5.031393
      p50 |    1.79615     1.3465     2.5698     1.9861     4.7917
      min |      .2717       .284       .489      .2593      1.943
      max |       6.16     6.5815     6.8563     6.0394    10.8275
       sd |   .9184741   .8846235   1.012532   .9231837   1.504337
```

A matched pair analysis (Wilcoxon sign rank test) was used to determine if a significant difference existed within individuals between the median volume of vessels on the right side and vessels on the left side. Significant differences were seen for the volume of the right versus left sigmoid sinus (p<0.001) as well as the median volume of the right versus left transverse sinus (p<0.001). For both the transverse and sigmoid sinuses the sinus on the right had a significantly larger volume than the sinus on the left. This is aligned with the majority of participants being right dominant (refer to section 4.4 below).

4.3.2 Venous sinus volume and sex

Significant differences were noted in the median volume of all vessels between males and females except for the left sigmoid sinus, as seen in the tabel 4.2 below. Vessels in males are shown to have greater volumes than the vessels of females, refer to table 4.3 below.

Table 4. 2 Difference in median volume of vessels between males and females, significant values highlighted in yellow.

Vessel	Median volume (cm^3)		p-value
	Male	Female	
Transverse (R)	2.138	1.559	<0.001
Transverse (L)	1.687	1.192	<0.001
Sigmoid (R)	2.878	2.458	0.0027
Sigmoid (L)	2.112	1.932	0.1050
Superior Sagittal Sinus	5.501	4.419	<0.001

Table 4. 3 Descriptive statistics showing the difference in the minimum, maximum and mean volumes of the cerebral venous sinuses between males and females.

Sex		Vol RT	Vol LT	Vol Sig R	Vol Sig L	Vol SSS
Female	N	179	179	178	177	180
	Mean	1.6812	1.3974	2.5448	2.0585	4.6213
	Min	0.2717	0.3992	0.489	0.6177	1.943
	Max	4.2939	4.202	5.7884	6.0394	9.2661
	SD	0.8074	0.7809	0.9574	0.8155	1.2598
Male	N	121	118	121	118	121
	Mean	2.27	1.7907	2.8974	2.2749	5.6414
	Min	0.4844	0.284	0.7959	0.2593	2.4844
	Max	6.16	6.5815	6.8563	5.1162	10.8275
	SD	0.9625	0.9789	1.058	1.0549	1.6317

It is generally accepted that males have a 10% larger brain in terms of weight than females (Dekaban, 1978). This difference in size was taken into account by decreasing the raw data of the males by 10% and the data was re-analyzed, refer to section 4.11.

4.3.3 Venous sinus volume and age

The SSS volume could not be measured for subject 083 (aged 22), as a complete view of the sinus could not be otained; and thus the subject was excluded from volume and dominance calculations.

The volume of the SSS was compared for stage of life (table 4.4) and age in ten-year intervals (table 4.3).

A Kruskal Wallis test indicated a significant difference in the median volume of the SSS between age groups (p= 0.0048), showing that the volume of the SSS is decreasing with age.

Spearman's rank correlation indicated no association between vessel volume and age for all vessels except the SSS. For the SSS a weak negative association (r_s = -0.1887, p = 0.0010) was noted.

Table 4. 4 Volume of the SSS between age groups (stage of life) (copied from statistical analyses report for the current study).

agecat	N	mean	p50	min	max	sd
0-18	14	6.146193	5.91285	4.4196	8.2055	1.124351
19-25	23	5.443135	5.0336	3.6418	9.2946	1.574788
26-55	172	5.019636	4.75205	2.3107	10.8275	1.594891
56-64	54	4.791774	4.7564	2.2565	8.7157	1.309557
65+	38	4.765192	4.63425	1.943	6.8191	1.218552
Total	301	5.031393	4.7917	1.943	10.8275	1.504337

Table 4. 5 Volume of the SSS between age groups (ten-year intervals) (copied from statistical analyses report for the current study).

agecat10s	N	mean	p50	min	max	sd
0-19	21	6.1815	5.904	4.3221	9.2946	1.425593
20-29	36	5.305228	5.02285	2.7349	9.2661	1.562906
30-39	54	5.075233	4.4995	2.3107	10.0089	1.728815
40-49	57	5.02514	4.8749	2.7073	10.8275	1.582977
50-59	66	4.699291	4.555	2.2565	8.112	1.277202
60+	67	4.820907	4.6995	1.943	8.7157	1.274311
Total	301	5.031393	4.7917	1.943	10.8275	1.504337

4.3.4 Venous sinus volume and pathology

Venous sinus volume was compared with the presence of cerebral pathology in the sample. Table 4.6 below shows the frequency cerebral pathology was seen in the sample.

Table 4. 6 Table showing the frequency of pathology within the sample.

Pathology		
	Frequency (N)	**Percent (%)**
Present	245	81.13
Absent	45	14.9
Unknown	12	3.97
Total	302	100

No significant difference was seen in the median volumes of the cerebral venous sinuses between patients with pathology and those without, see table 4.7 below.

Table 4. 7 Table showing the mean, minimum and maximum values for accessory vessels with and without pathology.

Pathology		R Transverse	L Transverse	R Sigmoid	L Sigmoid	SSS
Present	N	243	241	242	240	245
	Mean	1.9366	1.5788	2.6735	2.1513	5.0584
	Min	0.2717	0.3992	0.7959	0.2593	1.943
	Max	6.16	6.5815	6.8563	5.1162	10.8275
	SD	0.9394	0.9025	0.99712	0.9365	1.528
Absent	N	45	44	45	44	44
	Mean	1.7836	1.6025	2.4907	2.1875	4.8262
	Min	0.5804	0.284	0.489	0.6432	2.3107
	Max	3.8712	3.7914	4.87	6.0394	9.2661
	SD	0.822	0.8249	0.9581	0.9221	1.4755
Unknown	N	12	12	12	11	12
	Mean	2.0636	0.8704	3.7081	1.8401	5.2319
	Min	1.1483	0.355	2.3005	1.0407	2.9753
	Max	4.194	1.361	6.1539	3.0128	6.7847
	SD	0.8429	0.3064	1.01	0.5746	1.0876
Total	N	300	297	299	295	301
	Mean	1.9187	1.5537	2.6875	2.1451	5.0314
	Minimum	0.2717	0.284	0.489	0.2593	1.943
	Maximum	6.16	6.5815	6.8563	6.0394	10.8275
	SD	0.9185	0.8846	1.0125	0.9232	1.5043

No association was seen between the volume of the sinus and the presence or absence of pathology in the sample, refer to table 4.8 below.

Table 4. 8 Table showing the odds ratio, standard error and p-value for volume of the venous sinus compared with the presence of pathology.

	Odds Ratio	Std. Err.	*p-value*
Vol RT	0.971	0.3041	0.925
Vol LT	1.0183	0.3467	0.958
Vol R Sig	1.3031	0.3446	0.317
Vol L Sig	0.9223	0.2643	0.778
Vol SSS	1.0783	0.1766	0.645

The presence of cerebral pathology was also compared to the sinus volume within the individual (as opposed to the entire sample as seen in tables 4.7 and 4.8). No significant association was found withiin individuals between sinus volume and the presence of pathology, see table 4.9 below.

Table 4. 9 No significant association was found within individuals between the volume of the sinus and the presence of pathology.

Vessel	*p-value*
R Transverse	0.3441
L Transverse	0.6937
R Sigmoid	0.2005
L Sigmoid	0.7524
SSS	0.3041

4.3.5 Venous sinus volume and presence of accessory vessels

Accessory vessels were grouped as being present, absent or unknown (the researcher could not confidently identify the vessel on MR imaging in any of the image planes), see table 4.10 below.

Table 4. 10 Number of accessory vessels present, absent or unknown in the sample.

Accessory Vessel	Number Present	Number Absent	Unknown	Total
V. of Trolard *(Right)*	180	4	118	302
V. of Trolard *(Left)*	167	6	129	302
V. of Labbé *(Right)*	238	3	61	302
V. of Labbé *(Left)*	235	3	64	302
V. of Galen	299	0	3	302
Internal cerebral vv.	300	0	2	302

The median volumes of the venous sinuses were compared to the presence of the accessoray vessels. No significant association was seen between the median volumes of the sinuses and the right vein of Trolard.

A significant difference in the median volume of the right transverse sinus (p=0.0052) and the SSS (p=0.0137) was seen when the left vein of Trolard was either present or absent. In both cases the median volume of the sinuses were significantly lower when the vein of Trolard was absent. These results should be interpreted with caution due to the small sample size of cases where the left vein of Trolard was absent (N=6), see table 4.11 below.

Table 4. 11 Table showing median volume of the venous sinuses when compared to the presence of accessory vessels. Significant p-values are highlighted in yellow.

	Accessory Vessel			
	R v. of Trolard	L v. of Trolard	R v. of Labbé	L v. of Labbé
R Transverse	0.1173	0.0052	0.7263	0.5295
L Transverse	0.4896	0.1792	0.2645	0.1791
R Sigmoid	0.5064	0.3155	0.4237	0.9428
L Sigmoid	0.8921	0.5832	0.6594	0.5589
SSS	0.8429	0.0137	0.1021	0.3511

No significant difference was seen in the median volumes of the venous sinuses and the right or left veins of Labbé.

The influence of the vein of Galen and the internal cerebral veins on the median volume of the venous sinuses could not be tested as there were no confirmed cases for which they were absent.

4.3.6 Venous sinus volume and variation in the confluence of sinuses

Variation in the confluence of sinuses was clasified into three types (refer to table 4.12):

1) True confluence
2) Ring structure present within the confluence
3) No confluence

Table 4. 12 Table showing the number of persons per type of variation of the confluence of sinuses.

Type	Frequency (N)
1	126
2	78
3	98
Total	302

Variation in the confluence of sinuses had a significant effect on only the right and left sigmoid sinuses and not on the other sinuses. A significant difference was noted in the median volume of the right (p=0.0287) and left (p=0.0302) sigmoid sinuses, refer to table 4.13 below.

Further investigation using post hoc pairwise tests indicated that individuals with variation type 3 (no confluence) had a significantly increased volume in the right sigmoid sinus compared to those with no variation (the post hoc pairwise test gave a p value of 0.033). Post hoc pairwise tests did not indicate that any type of variation of the confluence significantly influenced the volume of the left sigmoid sinus more than the right.

Table 4. 13 Table showing the p-values for the association between variation in the confluence of sinuses and the volume of the venous sinuses.

Venous sinus	*p-value*
R Transverse	0.3033
L Transverse	0.3039
R Sigmoid	0.0287
L Sigmoid	0.0302
SSS	0.4415

4.3.7 Venous sinus volume and variation in the drainage pattern of the superior sagittal sinus

The variation in the drainage pattern of the SSS could be grouped into nine types, with type one being the generally acceptable configuration of the SSS, refer to table 4.14 below for the frequency each of the types were present in the sample.

Table 4. 14 Table showing the different types of variation of the SSS as well as the frequency of each type.

Type	Description	Frequency (N)
1	Generally acceptable anatomical configuration	112
2	The SSS drains only into the RT, with the SS draining into the RT and LT	78
3	The SSS drains only into the LT, with the SS draining into the RT and LT	34
4	The SSS bifurcates and drains into the RT and LT respectively and not into the confluence of sinuses.	54
5	The SSS drains into the right transverse sinuses and the SS drains into the left transverse sinus (SSS to the RT and the SS to the LT)	9
6	The SSS drains into the left transverse sinuses and the SS drains into the right transverse sinus (SSS to the LT and the SS to the RT)	4
7	The SSS drains to both the RT and LT while the SS drains only to the LT	6
8	The SSS drains to both the RT and LT while the SS drains only to the RT	4
9	The SSS and SS drain to the same side.*	1

** Variation type nine only had one case, in which. Both the SSS and SS drained into the right transverse sinus.*

No significant difference was seen in the volume of the SSS between the different variation types (p=0.4154). Thus variation of the SSS did not significantly influence the volume of the SSS.

The volumes of each of the other sinuses were also compared with the different variation types of the SSS. Significant differences in the median volumes of the right transverse sinus (p<0.001), left transverse sinus (p<0.001), right sigmoid sinus (p<0.001) and left sigmoid sinus (p<0.001) were seen between the different variation types of the SSS.

Table 4.15 below illustrates for which type of variation a significant difference in the volume of the sinus was seen. No significant difference was seen in the sinuse volumes when compared to variation types 5 – 9, it should be noted that the number of patients showing variation types 5 – 9 was extremely small and thus a significant difference could be missed in the analyses due to this sample size.

Table 4. 15 Table showing the significant median vessel volumes with the corresponding variation type.

Vessel and SSS variation type	Median volume (cm^3)	p-value*
Transverse (R)		
No variation	2.062	
Type 3	1.395	0.002
Type 4	1.356	0.003
Transverse (L)		
No variation	1.495	
Type 2	0.94	<0.001
Sigmoid (R)		
No variation	2.482	
Type 2	3.067	<0.001
Sigmoid (L)		
No variation	2.01	
Type 2	1.647	0.014
Type 3	2.694	0.041
* pairwise comparison to median vessel volume with no variation in the SSS present. Bonferonni correction applied		

4.3.8 Pooled venous volume on the right and left sides

The volumes for the right and left sides were pooled as follows. The volume of the right transverse sinus was added to the right sigmoid sinus and same was done for the left transverse and left sigmoid sinuses. This combined volume was then further analysed.

The combined volume on the left and right sides showed no significant association with age. However, when the pooled right and pooled left volumes were compared between the sexes, males were shown to have significantly larger vessels (p=0.0016) table 4.16 below.

Table 4. 16 Descriptive statistics on the pooled volumes for the sinuses on the left and right respectively.

Sex		Left pooled	Right pooled
Female	N	176	177
	Mean	3.4682	4.2304
	Min	1.0473	1.324
	Max	8.8308	8.9922
	SD	1.452	1.5416
Male	N	115	121
	Mean	4.1347	5.1674
	Min	0.8254	1.5045
	Max	11.2344	11.7597
	SD	1.848	1.8015

No association was seen between the combined volumes on the left (p=0.7441) or the right (p=0.3633) and the presence of pathology.

No association was seen between the combined volumes on the left (p=0.1185) or the right (p=0.0854) and variation of the confluence of sinuses.

The combined volumes on the left and right were compared with the different types of drainage patterns of the SSS. The analysis of pooled left and right volumes did not result in different results from those seen in table 4.15 above.

4.4 Circulatory dominance determined by volume of venous blood

Building on the work by Kitamura *et al.,* (2017), a sliding scale was constructed for the current study to visualise circulatory dominance in the venous system as a continuum rather than separate data points (figure 4.6).

Dominance was expressed as a ratio of the right to left transverse sinuses (thus, the sinus was more than 50% than that of the contralateral side). If the ratio was more than 1.5, the sinus was classified as right dominant. If the ratio was less than 0.67, it was considered left dominant. If the ratio fell between 1.5 and 0.67 the sinuses were considered to be equally dominant (co-dominant).

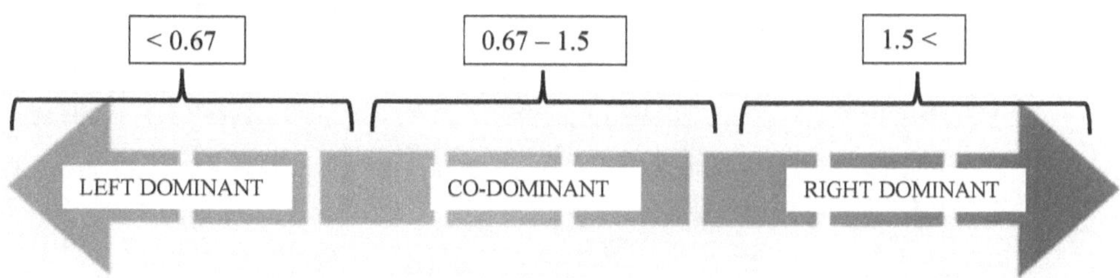

Figure 4.6: Dominance ratio displayed on a sliding scale.

The majority of individuals in the study displayed right side dominance (41.72%), followed by co-dominance (39.40%) and left side dominance (18.87%), refer to table 4.17.

Table 4. 17 Frequency of dominance within the study sample (copied from statistical analyses report for the current study).

```
 dominance |      Freq.       Percent
-----------+---------------------------
Co-dominant |       119         39.40
      Left |        57         18.87
     Right |       126         41.72
-----------+---------------------------
     Total |       302        100.00
```

There is no significant difference in the type of circulatory dominance displayed between males and females (p=0.686), table 4.18.

Table 4. 18 Table showing the distribution of dominance between the sexes.

	Female	Male	Total
Co-Dominant	73	46	119
Left Dominant	36	21	57
Right Dominant	72	54	126
Total	181	121	302

There is no significant difference in the median age associated with each type of dominance, refer to tables 4.19 (p=0.570), and 4.20 (p=0.454) below.

Table 4. 19 Table showing the distribution of dominance for age (stage of life).

	0-18	19-25	26-55	56-64	65+	Total
Co-Dominant	4	11	68	19	17	119
Left Dominant	3	5	37	6	6	57
Right Dominant	7	8	67	29	15	126

Table 4. 20 Table showing the distribution of dominance for age (ten-year intervals).

	0-19	20-29	30-39	40-49	50-59	60+	Total
Co-Dominant	9	12	27	21	20	30	119
Left Dominant	5	7	11	12	14	8	57
Right Dominant	7	18	16	24	32	29	126

There is no significant association in the type of circulatory dominance and presence of pathology (p=0.998), table 4.21.

Table 4. 21 Table showing distribution of pathology compared with circulatory dominance (copied from statistical analyses report for the current study).

```
                |       path
   dominance    |    No      Yes  |    Total
 ---------------+-----------------------+----------
   Co-dominant  |    18       99  |      117
         Left   |     9       48  |       57
        Right   |    18       98  |      116
 ---------------+-----------------------+----------
        Total   |    45      245  |      290
```

The presence or absence of accessory vessels was not associated with any type of circulatory dominance, refer to table 4.22 below. For further analyses refer to section 4.6.

Table 4. 22 Table showing the p-value for the association between dominance and the presence or absence of accessory vessels.

Accessory vessel	*p-value*
R v. of Trolard	0.828
L v. of Trolard	0.282
R v. of Labbé	0.683
L v. of Labbé	0.782
Internal cerebral vv.	0.339

For all patients for which dominance was determined the vein of Galen was present, therefore the presence or the absence of the vein of Galen could not be associated with dominance.

For the association between dominance and variations of the SSS and confluence of sinuses see sections 4.7 and 4.8.

4.5 Hypoplasia and aplasia

Shapiro Wilk tests were used to determine if numerical variables were normally distributed. Neither age nor any of the volumes of vessels were normally distributed and therefore non-parametric tests were conducted.

Differences within numerical variables were assessed using the Wilcoxon Rank Sum test (for two group comparisons) and the Kruskal Wallis test (for multi-group comparisons). Where necessary pairwise post-hoc testing was conducted with Bonferonni correction applied.

Categorical variables were analysed using the Pearson's Chi-squared test. For matched pair analysis of data for the left and right sides, a McNemar chi-square test was used. Two sample tests of proportions were used to determine significant differences between pairwise proportions where appropriate.

4.5.1 Hypoplasia of the right and left transverse sinuses

Data was available for 296 individuals on the left side and 299 individuals on the right side. For the matched analysis, 294 individuals had data for both the left and right sides.

The majority of individuals in the study displayed hypoplasia, namely 234 (table 4.23) on the right side 266 (table 4.24) on the left side.

Table 4. 23 Frequency of hypoplasia in the right transverse sinus (copied from statistical analyses report for the current study).

```
Hypoplastic |
         RT |        Freq.        Percent
------------+------------------------------
         No |           65          21.74
        Yes |          234          78.26
------------+------------------------------
      Total |          299         100.00
```

Table 4. 24 Frequency of hypoplasia in the left transverse sinus (copied from statistical analyses report for the current study).

```
Hypoplastic |
         LT |        Freq.        Percent
------------+------------------------------
         No |           30          10.14
        Yes |          266          89.86
------------+------------------------------
      Total |          296         100.00
```

Very few patients showed hypoplasia only on the one side. Fifty-seven patients showed hypoplasia only on the left, whereas, 25 showed hypoplasia only on the right, refer to graph 4.5 and table 4.25 below.

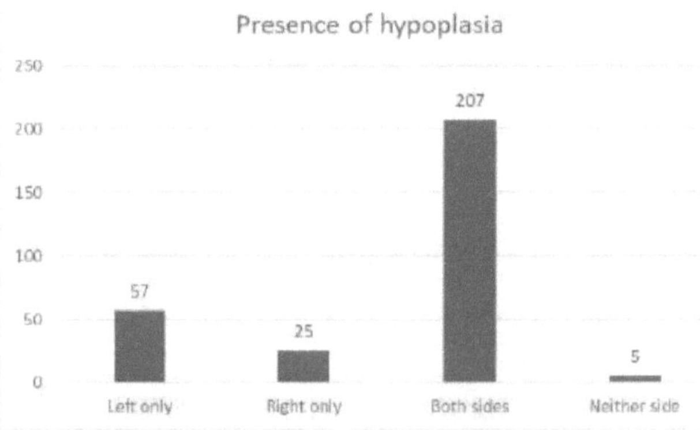

Graph 4. 3 Presence of hypoplasia in the current sample.

Table 4. 25 Frequency (N) of hypoplasia in the right and left transverse sinuses (copied from statistical analyses report for the current study).

```
Hypoplasti |      Hypoplastic LT
       c RT |       No         Yes |
------------+-----------------------+-
        No |        5          57 |
       Yes |       25         207 |
------------+-----------------------+-
     Total |       30         264 |
```

The matched pair analyses indicated a significant association between hypoplasia on the left- and right-hand side (p= 0.0004). The majority of individuals displayed hypoplasia on both sides (207; 70.41%). Therefore if a patient displayed hypoplasia on the one side there was a significant association with hypoplasia on the other side.

Significant differences were noted in the median volume of vessels and the presence of hypoplasia on the right and left sides (see table 4.26 below). Typically, the presence of hypoplasia on the left side was associated with a lower volume of the left transverse and sigmoid sinuses and a greater volume in the right sigmoid sinus. The presence of hypoplasia on the right side was associated with a lower volume of the right transverse and sigmoid sinuses and a greater volume in the left sigmoid sinus.

Table 4. 26 Significant difference in the median volume of vessels and the presence of hypoplasia on the left and right sides.

Vessel volume	Median volume (cm3)		p-value
	Hypoplasia Absent	Hypoplasia Present	
Left side hypoplasia			
Transverse (R)	1.70	1.79	0.6594
Transverse (L)	2.72	1.25	<0.0001
Sigmoid (R)	2.03	2.66	0.0001
Sigmoid (L)	2.71	1.94	0.0002
Superior Sagittal Sinus	4.6	4.84	0.5525
Right side hypoplasia			
Transverse (R)	2.77	1.56	<0.0001
Transverse (L)	1.25	1.45	0.1658
Sigmoid (R)	3.37	2.43	<0.0001
Sigmoid (L)	1.56	2.12	<0.0001
Superior Sagittal Sinus	4.5	4.86	0.1006

4.5.2 Hypoplasia and sex

A significant association was found between the sex of an individual and hypoplasia on the right side (p=0.006), see table 4.27 below. Significantly more females displayed hypoplasia on the right side than did males. No significant association was seen between sex and hypoplasia on the left (p=0.422), see table 4.28 below.

Table 4. 27 Frequency (N) of hypoplasia of the right transverse sinus between males and females.

Hypoplastic RT	Sex	
	Female	Male
No	29	36
Yes	149	85
Total	178	121

Table 4. 28 Frequency (N) of the hypoplasia of the left transverse sinus between males and females.

Hypoplastic LT	Sex	
	Female	Male
No	16	14
Yes	162	104
Total	178	118

4.5.3 Hypoplasia and age

There was no significant difference in the median age of individuals with or without hypoplasia on the right side (p=0.0715) and the left side (p=0.7280), refer to figures 4.7 and 4.8 below.

```
Ho: Age(Hypop~RT==No) = Age(Hypop~RT==Yes)
             z =    1.802
    Prob > |z| =    0.0715
```

Figure 4.7: The median age of individuals and hypoplasia on the right (copied from statistical analyses report for the current study).

```
Ho: Age(Hypop~LT==No) = Age(Hypop~LT==Yes)
             z =   -0.348
    Prob > |z| =    0.7280
```

Figure 4.8: The median age of individuals and hypoplasia on the left (copied from statistical analyses report for the current study).

No association was found between age for stage of life (table 4.29) or ten-year intervals (table 4.30) and the presence of hypoplasia on the right side.

Table 4. 29 Presence of a hypoplastic right transverse sinus in terms of stage of life.

Hypoplastic RT	Age in terms of stage of life					
	0 - 18	19 - 25	26 - 55	56 - 64	65+	Total
No	1	5	35	13	11	65
Yes	14	17	135	41	27	234
Total	15	22	170	54	38	299

Table 4. 30 Presence of a hypoplastic right transverse sinus in 10-year intervals.

Hypoplasia RT	Age in 10-year intervals					
	0 - 19	20 - 29	30 - 29	40 - 49	50 - 59	60+	Total
No	2	6	12	12	14	19	65
Yes	20	29	42	44	51	48	234
Total	22	35	54	56	65	67	299

No association was found between age for stage of life (table 4.31) or ten-year intervals (table 4.32) and the presence of hypoplasia on the left side.

Table 4. 31 Presence of a hypoplastic left transverse sinus in terms of stage of life.

Hypoplastic LT	Age in terms of stage of life					
	0 - 18	19 - 25	26 - 55	56 - 64	65+	Total
No	3	4	15	2	6	30
Yes	12	18	153	51	32	266
Total	15	22	168	53	38	299

Table 4. 32 Presence of a hypoplastic left transverse sinus in 10-year intervals.

Hypoplasia LT	Age in 10-year intervals					
	0 - 19	20 - 29	30 - 29	40 - 49	50 - 59	60+	Total
No	5	2	5	5	5	8	30
Yes	17	33	47	51	59	59	266
Total	22	35	52	56	64	67	296

4.5.4 Hypoplasia and dominance

Frequency of hypoplasia on the left (table 4.33) and right (table 4.34) were compared to the different types of dominance seen in the sample.

Table 4. 33 Dominance and hypoplasia of the left transverse sinus (copied from statistical analyses report for the current study).

```
              | Hypoplastic  LT
 Dominance    |    No        Yes  |     Total
--------------+-------------------+----------
Co-dominance  |    11        108  |      119
       Left   |    19         38  |       57
      Right   |     0        120  |      120
--------------+-------------------+----------
      Total   |    30        266  |      296
```

Table 4. 34 Dominance and hypoplasia of the right transverse sinus (copied from statistical analyses report for the current study).

```
              | Hypoplastic  RT
 Dominance    |    No        Yes  |     Total
--------------+-------------------+----------
Co-dominance  |    12        107  |      119
       Left   |     0         55  |       55
      Right   |    53         72  |      125
--------------+-------------------+----------
      Total   |    65        234  |      299
```

A significant association was found between the type of dominance and hypoplasia on the right side (p<0.001) and the left side (p<0.001), refer to table 4.35 below. Pairwise tests indicated that left side hypoplasia is significantly more common amongst individuals with right side circulatory dominance. Similarly, right side hypoplasia is significantly more common amongst individuals with left side circulatory dominance.

Table 4. 35 Association between dominance and hypoplasia of the left and right transverse sinuses respectively.

Circulatory dominance	Absent	Present	p-value
Left hand side hypoplasia			
Co-dominance	11 (36.67%)	108 (40.6%)	0.6769
Left dominance	19 (63.3%)	38 (14.29%)	< 0.0001
Right dominance	0	120 (45.11%)	< 0.0001
Right hand side hypoplasia			
Co-dominance	12 (18.46%)	107 (45.72%)	= 0.0001
Left dominance	0	55 (23.5%)	< 0.0001
Right dominance	53 (81.53%)	72 (30.77%)	< 0.0001

4.5.5 Hypoplasia and pathology

The presence and absence of pathology was compared to the presence of hypoplasia on the left (table 4.36) and the right (table 4.37).

No significant association was found between the presence of pathology and hypoplasia on the left side (p=0.775) or the right side (p=0.943).

Table 4. 36 Frequency (N) of pathology and presence of hypoplasia of the left transverse sinus.

	Hypoplastic LT	
Pathology	**No**	**Yes**
Yes	26	215
No	4	39
Unknown	0	12
Total	30	266

Table 4. 37 Frequency (N) of pathology and presence of hypoplasia of the right transverse sinus.

	Hypoplastic RT	
Pathology	**No**	**Yes**
Yes	52	191
No	10	34
Unknown	3	9
Total	65	234

4.5.6 Hypoplasia and presence of accessory vessels

The presence or absence of accessory vessels was not associated with the presence of hypoplasia on the right side (table 4.38) or the left side (table 4.39).

Table 4. 38 Association between the hypoplasia of the right transverse sinus and presence of accessory vessels.

Accessory vessel	**p- value**
Right v. of Trolard	0.669
Left v. of Trolard	0.539
Right v. of Labbé	0.079
Left v. of Labbé	0.115
Vein of Galen	0.625
Internal cerebral vv.	0.455

Table 4. 39 Association between the hypoplasia of the left transverse sinus and presence of accessory vessels.

Accessory vessel	p- value
Right v. of Trolard	0.607
Left v. of Trolard	0.843
Right v. of Labbé	0.714
Left v. of Labbé	0.815
Vein of Galen	0.181
Internal cerebral vv.	0.634

4.5.7 Hypoplasia and variation of the confluence of sinuses

No significant association was found between variation within the confluence of sinuses and the presence of hypoplasia on the left side (p=0.952) or on the right side (p=0.635).

The number of patients with hypoplasia on the left (table 4.40) and on the right (table 4.41) compared to the types of variation of the confluence of sinuses is seen below.

Table 4. 40 Frequency (N) of the different variation types in the confluence of sinuses and hypoplasia of the left transverse sinus.

Variation type	Hypoplastic LT	
	No	Yes
1	13	109
2	8	70
3	9	87
Total	30	266

Table 4. 41 Frequency (N) of the different variation types in the confluence of sinuses and hypoplasia of the right transverse sinus.

Variation type	Hypoplastic RT	
	No	Yes
1	28	95
2	14	64
3	23	75
Total	65	234

4.5.8. Hypoplasia and variation in the pattern of drainage of the superior sagittal sinus

A significant association was found between variation of the SSS and hypoplasia of the left transverse sinus (p=0.019), see table 4.42 below. Further analysis indicated variation type 2 was significantly associated with hypoplasia on the left (p=0.0108) and variation type 8 was significantly associated with no hypoplasia on the left (p = 0.0078).

Table 4. 42 Significant association between variation of the SSS and hypoplasia of the left transverse sinus (copied from statistical analyses report for the current study).

```
Variation  |   Hypoplastic  LT
  of SSS   |      No        Yes  |      Total
-----------+---------------------+-----------
        1  |      13         96  |       109
        2  |       2         75  |        77
        3  |       4         30  |        34
        4  |       7         47  |        54
        5  |       0          9  |         9
        6  |       1          3  |         4
        7  |       1          4  |         5
        8  |       2          2  |         4
-----------+---------------------+-----------
    Total  |      30        266  |       296

           Pearson chi2(7)  =   15.2599    Pr =  0.033
           Fisher's exact   =                    0.019
```

Hypoplasia of the right transverse sinus was significantly associated with variation of the SSS (p = 0.004) see table 4.43 below. Further analysis indicated variation type 2 was significantly associated with no hypoplasia on the right (p=0.0039), while variation types 3 and 4 were significantly associated with hypoplasia on the right (p=0.0173 and p=0.014 respectively).

Table 4. 43 Significant association between variation of the SSS and hypoplasia of the right transverse sinus (copied from statistical analyses report for the current study).

```
Variation  |   Hypoplastic  RT
  of SSS   |      No        Yes  |      Total
-----------+---------------------+-----------
        1  |      28         82  |       110
        2  |      26         52  |        78
        3  |       2         32  |        34
        4  |       5         49  |        54
        5  |       3          6  |         9
        6  |       1          3  |         4
        7  |       0          5  |         5
        8  |       0          4  |         4
        9  |       0          1  |         1
-----------+---------------------+-----------
    Total  |      65        234  |       299

           Pearson chi2(8)  =   20.5376    Pr =  0.008
           Fisher's exact   =                    0.004
```

<u>**4.6 Accessory vessels**</u>

The results of this analyses should be interpreted with caution due to the small number of cases where the veins were seen as absent.

For the analysis below, in situations where it could not be accurately determined if the vessel was present or absent the case was marked as unknown and excluded from the sample analysis, see table 4.44 below.

Table 4. 44 Table showing the present and absent cases for the accessory vessels.

	Present	**Absent**	**Total**
Right v. of Trolard	181	4	302
Left v. of Trolard	168	6	302
Right v. of Labbé	239	4	302
Left v. of Labbé	236	3	302
V. of Galen	299	0	302
Internal cerebral vv.	300	0	302

There is no significant difference in the presence of accessory vessels between males and females.

Age does not play a role in the presence/ absence of a vessel and was therefore not analysed.

A Chi-square test was done to determine whether there was a statistically significant difference between the expected frequencies and the observed frequencies. No significant association was found between the presence of any accessory vessel and the presence of pathology. The assessment was repeated using Fischer's exact test (as the study has a relatively smaller sample size and the significance of the deviation can be calculated exactly and is not an approximation as with a chi-square test) still no significant association was found.

The relationship between left, right and co-dominance was compared with presence of each accessory vessel respectively. It was found that the presence or absence of the right vein of Trolard (table 4.45) and the left vein of Trolard (table 4.46) was not associated with any type of dominance.

Table 4. 45 Presence of the right vein of Trolard and dominance (copied from statistical analyses report for the current study).

```
                     |        trolardr
       dominance |    Absent      Present |       Total
      ---------------+-----------------------------+-----------
      Co-dominant |         1           73 |          74
                     |       1.6         72.4 |        74.0
      ---------------+-----------------------------+-----------
             Left |         1           32 |          33
                     |       0.7         32.3 |        33.0
      ---------------+-----------------------------+-----------
            Right |         2           76 |          78
                     |       1.7         76.3 |        78.0
      ---------------+-----------------------------+-----------
            Total |         4          181 |         185
                     |       4.0        181.0 |       185.0

           Fisher's exact =                         0.828
```

Table 4. 46 Presence of the left vein of Trolard and dominance (copied from statistical analyses report for the current study).

```
             |       trolardl
  dominance  |    Absent    Present  |      Total
-------------+------------------------+-----------
Co-dominant  |         4         64  |         68
             |       2.3       65.7  |       68.0
-------------+------------------------+-----------
       Left  |         1         29  |         30
             |       1.0       29.0  |       30.0
-------------+------------------------+-----------
      Right  |         1         75  |         76
             |       2.6       73.4  |       76.0
-------------+------------------------+-----------
      Total  |         6        168  |        174
             |       6.0      168.0  |      174.0

         Fisher's exact =                  0.282
```

The presence or absence of the right vein of Labbé (table 4.47) and the left vein of Labbé (table 4.48) was not associated with any type of dominance.

Table 4. 47 Presence of the right vein of Labbé and dominance (copied from statistical analyses report for the current study).

```
             |       labber
  dominance  |    Absent    Present  |      Total
-------------+------------------------+-----------
Co-dominant  |         1         99  |        100
             |       1.6       98.4  |      100.0
-------------+------------------------+-----------
       Left  |         1         46  |         47
             |       0.8       46.2  |       47.0
-------------+------------------------+-----------
      Right  |         2         94  |         96
             |       1.6       94.4  |       96.0
-------------+------------------------+-----------
      Total  |         4        239  |        243
             |       4.0      239.0  |      243.0

         Fisher's exact =                  0.683
```

Table 4. 48 Presence of the left vein of Labbé and dominance (copied from statistical analyses report for the current study).

```
                 |        labbel
    dominance |   Absent     Present |      Total
 -------------+------------------------+-----------
 Co-dominant  |        2          93 |         95
              |      1.2        93.8 |       95.0
 -------------+------------------------+-----------
         Left |        0          42 |         42
              |      0.5        41.5 |       42.0
 -------------+------------------------+-----------
        Right |        1         101 |        102
              |      1.3       100.7 |      102.0
 -------------+------------------------+-----------
        Total |        3         236 |        239

              |      3.0       236.0 |      239.0

        Fisher's exact =                    0.782
```

The vein of Galen was not included in testing as the vein of Galen was present in all cases for which dominance of the patient was known.

The presence or absence of the internal cerebral veins were not associated with any type of dominance, see table 4.49 below.

Table 4. 49 Presence of the internal cerebral veins and dominance (copied from statistical analyses report for the current study).

```
    dominance |   Present     Unknown |      Total
 -------------+------------------------+-----------
 Co-dominant  |      117           2 |        119
              |    118.2         0.8 |      119.0
 -------------+------------------------+-----------
         Left |       57           0 |         57
              |     56.6         0.4 |       57.0
 -------------+------------------------+-----------
        Right |      126           0 |        126
              |    125.2         0.8 |      126.0
 -------------+------------------------+-----------
        Total |      300           2 |        302
              |    300.0         2.0 |      302.0

        Fisher's exact =                    0.339
```

The presence or absence of accessory vessels was not associated with any variations in the SSS or variations within the confluence.

<u>**4.7 Variation within the drainage pattern of the superior sagittal sinus**</u>

The majority of individuals had a type 1 classification of the SSS (37.09%). Followed by type 2 (25.83%), type 4 (17.88%) and type 3 (11.26%) in order of decreasing frequency, refer to graph 4.7 below.

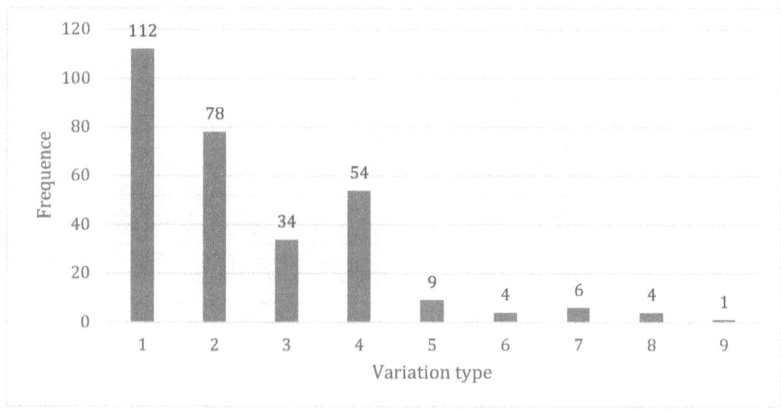

Graph 4. 4 Graph showing the frequency of the different types of variation in the drainage pattern of the SSS.

There is no significant association between males and females and variations of the SSS (p=0.489).

There is no significant association between age in terms of stage of life (table 4.50; p=0.120) or age in ten-year intervals (table 4.51; p=0.459) and variations of the SSS.

Table 4. 50 Frequency (N) of type of variation in the drainage pattern of the SSS according to stage of life.

Variation type	Age in terms of stage of life				
	0-18	**19-25**	**26-55**	**56-64**	**65+**
1	8	12	66	14	12
2	4	2	41	18	13
3	1	3	20	6	4
4	1	4	31	11	7
5	0	0	6	3	0
6	0	0	4	0	0
7	0	3	1	2	0
8	0	0	2	0	2
9	0	0	1	0	0

Table 4. 51 Frequency (N) of type of variation in the drainage pattern of the SSS according to ten-year intervals.

Variation type	Age in ten-year intervals					
	0-19	20-29	30-39	40-49	50-59	60+
1	11	14	21	18	29	19
2	5	7	15	14	14	23
3	2	5	2	9	7	9
4	3	7	11	9	13	11
5	0	0	2	4	1	2
6	0	1	2	1	0	0
7	0	3	0	0	2	1
8	0	0	1	1	0	2
9	0	0	0	1	0	0

There is no significant association with variations of the SSS and the presence of pathology (p=0.491).

There is a significant association with variation of the SSS and the type of dominance (p<0.001). Variation type 2 appears to be associated with right side dominance and variation type 4 appears to be associated with co-dominance. This is in line with the findings of increased volume in the right sigmoid sinus in conjunction with type 2 variation.

4.7.1 Examples of nine variation types in the drainage pattern of the superior sagittal sinus

For every subject a three-dimensional model was constructed of the dural venous sinuses. These models were studied and the gross anatomical structure documented.

The generally accepted drainage pattern of the SSS is shown in figure 4.7 below. The SSS drains into the confluence of sinuses as one vessel along with the transverse and SS.

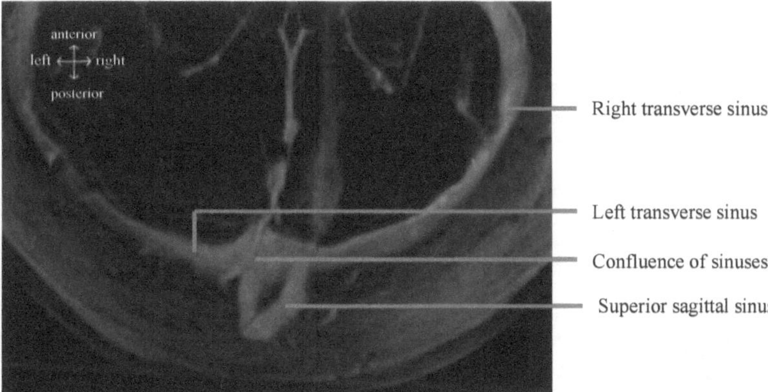

Figure 4.7: The superior sagittal, transverse and SS all drain into the confluence of sinus, type 1 drainage pattern. Image from the current research project.

In some instances the SSS drained only into the right transverse sinus, while the SS drained into both the left and right transverse sinuses. In figure 4.8 (A and B) below, the SSS is seen off centre and draining to the right transverse sinus and the SS is seen draining to both transverse sinuses (white lines). This was classified as a type 2 drainage pattern (78 cases).

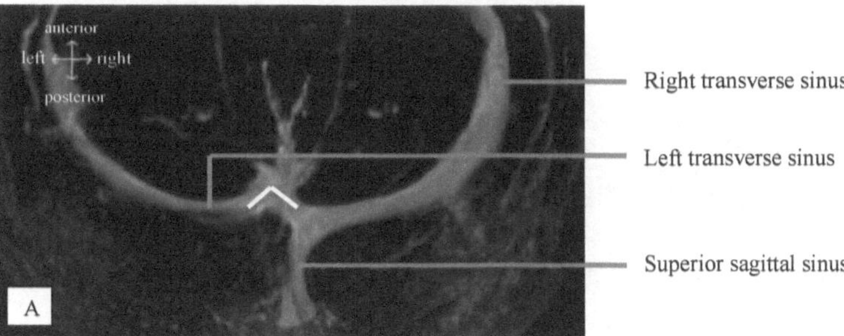

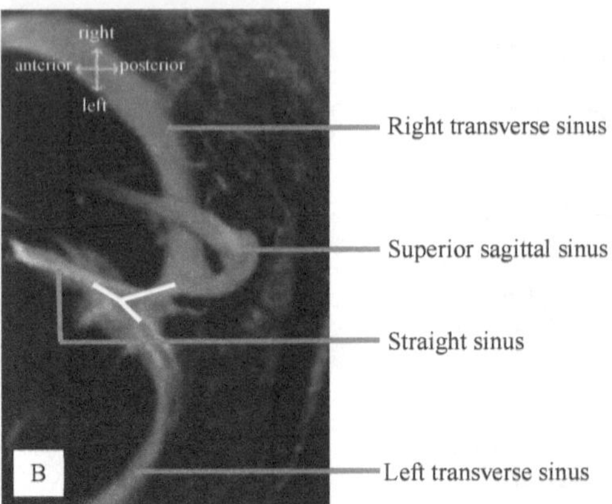

Figure 4.8: The SSS drains only to the right transverse sinus and the SS drains to both the transverse sinuses. **A:** Superior – inferior view of the drainage pattern of the SSS. **B:** Oblique view of the drainage pattern of the SSS. White lines – drainage of the SS. Image from the current research project.

The SSS was also seen only draining into the left transverse sinus and the SS draining into both the left and right transverse sinuses. This was referred to as a type 3 drainage pattern of the SSS (34 cases) it was seen in fewer cases than in type 2 described above (figure 4.9).

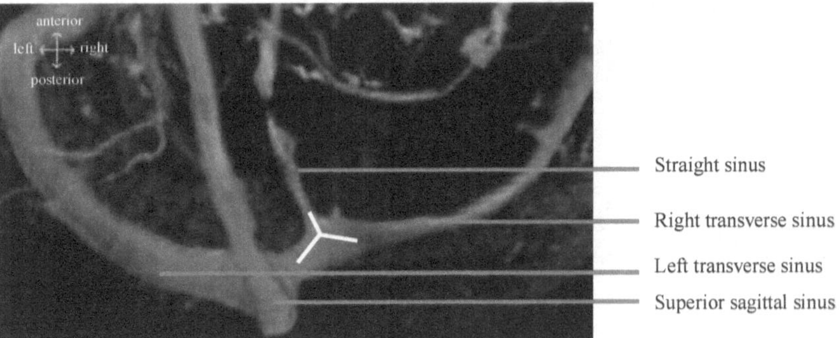

Figure 4.9: Type 3 drainage pattern of the SSS. The SSS can be seen draining only to the left transverse, while the SS is seen draining to both the transverse sinuses (white lines). Image from the current research project.

A type 4 drainage pattern was seen when the SSS bifurcates and drains into both the left and right transverse sinuses, the SS is also seen draining to both the transverse sinuses (54 cases). This drainage pattern of the SSS corresponds to variation type 2 of the confluence of sinuses. A ring-structure can be seen in the area where the confluence of sinuses is generally seen (figure 4.10).

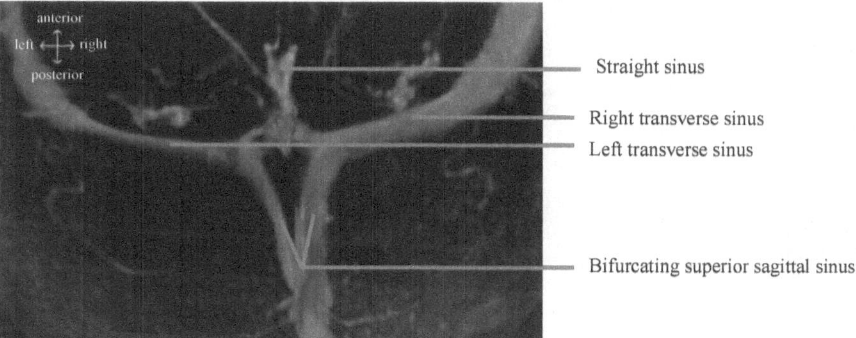

Figure 4.10: Type 4 drainage pattern of the SSS. The SSS bifurcates before draining into the left and right sinuses respectively. Image from the current research project.

In rare cases the SSS drained to the one transverse sinus and the SS drained to the other, with no connection between them. In nine cases the SSS drained to the right transverse and the straight drained to the left transverse, this was referred to as a type 5 drainage pattern (figure 4.11).

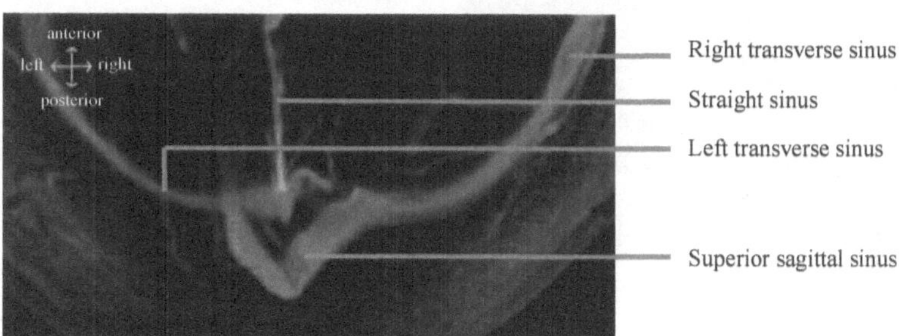

Figure 4.11: Type 5 drainage pattern of the SSS. No communication can be seen between the right and left transverse sinuses. Image from the current research project.

In four cases the superior sagittal drained to the left transverse and the straight to the right, thus a type 6 drainage pattern was seen (figure 4.12).

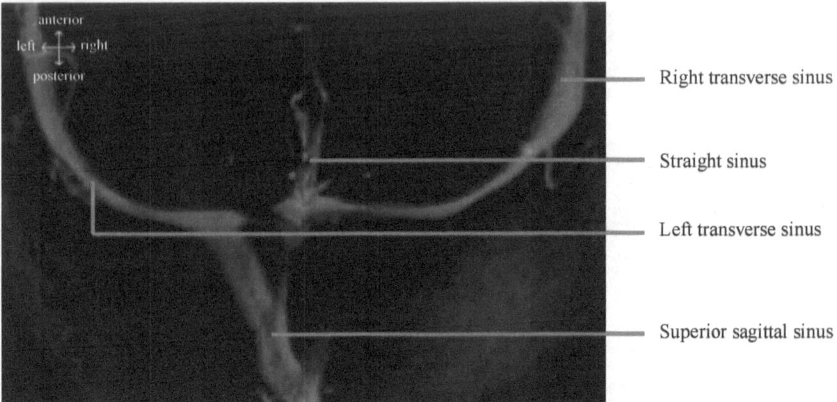

Figure 4.12: Type 6 drainage pattern of the SSS. No communication can be seen between the right and left transverse sinuses. Image from the current research project.

In six cases the SSS bifurcated before draining into both the transverse sinuses. The SS only drained to the left transverse sinus (figure 4.13).

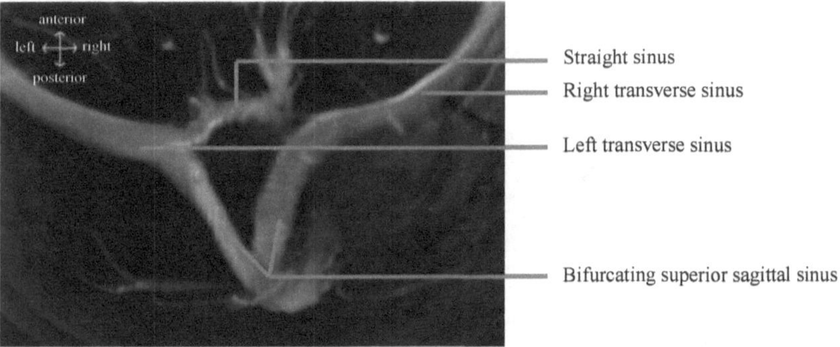

Figure 4.13: Type 7 drainage pattern. The SSS bifurcates and drains to both the transverse sinuses, the SS only drains to the left transverse sinus. Image from the current research project.

In four cases the SSS bifurcated before draining to both the transverse sinuses. The SS only drained to the right transverse sinus (figure 4.14).

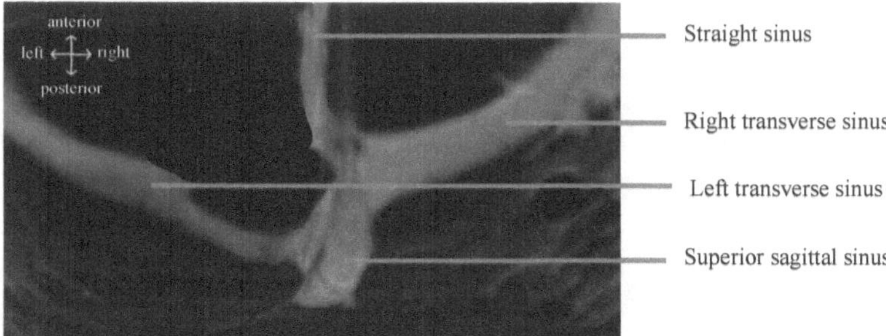

Figure 4.14: Type 8 drainage pattern. The SSS bifurcates and drains to both the transverse sinuses, the SS only drains to the right transverse sinus. Image from the current research project.

Only one case was found where the SSS and the SS drained into the same transverse sinus. In this case the contralateral transverse sinus was absent.

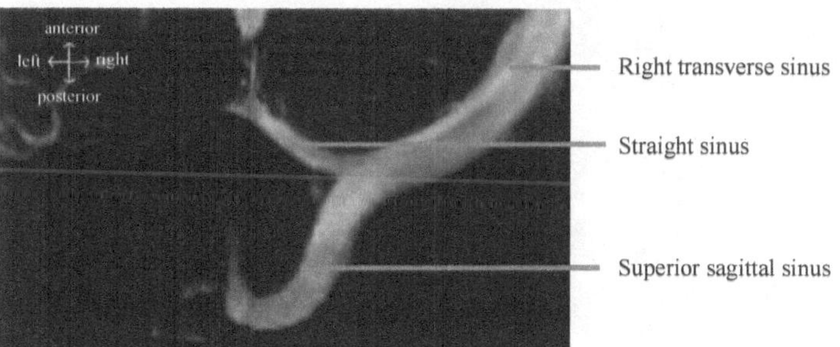

Figure 4.15: Type 9 drainage pattern. The superior sagittal and SS both drain into the right transverse sinus and the left transverse sinus is absent. Image from the current research project.

4.8 Variation within the confluence of sinuses

The majority of individuals had type 1 classification of the confluence of sinuses (41.72%). The most common variations present were type 3 (32.45%) followed by type 2 (25.83%), refer to graph 4.8 below.

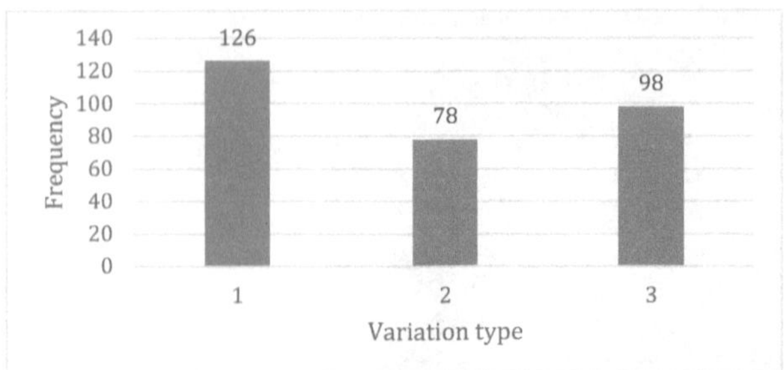

Graph 4. 5 Graph showing the frequency of the three types of variation of the confluence of sinuses.

A true confluence, meaning the superior sagittal, left and right transverse and the SS drain into the confluence of sinuses was classified as a type 1 confluence of sinuses, see figure 4.16 below.

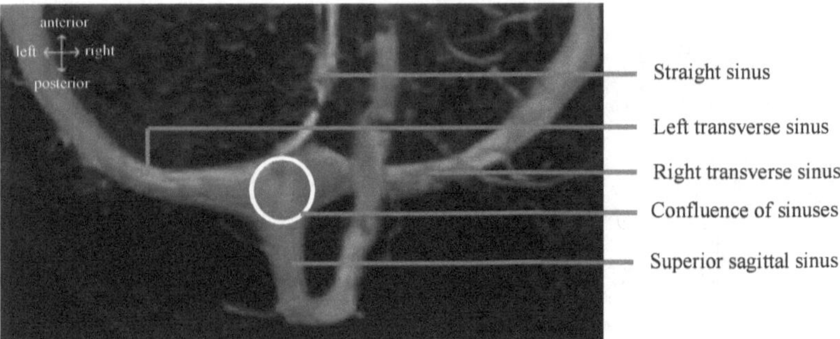

Figure 4.16: Type 1 variation of the confluence of sinuses.

A type 2 confluence of sinuses was seen when a ring structure was visible within the confluence. A ring structure was formed when the SSS and SS bifurcated and drained into the transverse sinuses, leaving a "gap" where the confluence would traditionally have been, see figure 4.17 below.

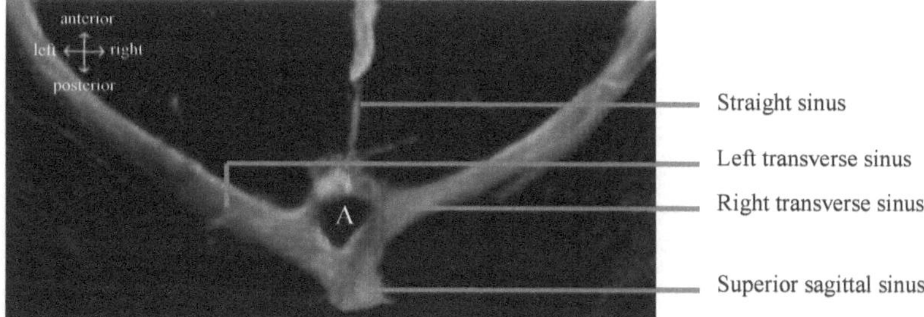

Figure 4.17: Type 2 variations of the confluence of sinuses. **A** = an area of loss of signal due to the ring-shaped connections between the dural venous sinuses in place of the usual confluence of sinuses.

In cases where there was no true confluence or where there was no clear communication between the different parts of the confluence (as in type 2) it was classified as non-confluence (type 3), see figure 4.18 below.

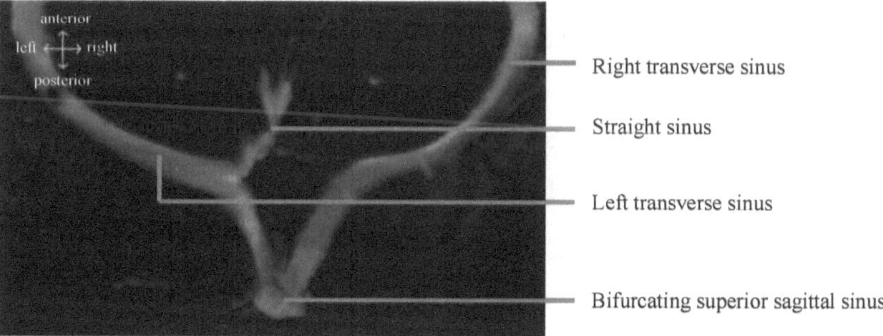

Figure 4.18: Type 3 variation of the confluence of sinuses.

There is no significant association between sex and variations within the confluence of sinuses (p=0.116), see table 4.52 below.

Table 4. 52 Frequency (N) of the different variation types of the confluence of sinuses for the sexes.

Variation type	Sex	
	Female	Male
1	83	43
2	40	38
3	58	40

There is no significant association between variations within the confluence of sinuses and age in terms of stage of life (table 4.53; p=0.343) or age in ten-year intervals (table 4.54; p=0.089).

Table 4. 53 Frequency (N) of the variation type of the confluence of sinuses for age in terms of stage of life.

Variation type	Age in terms of stage of life				
	0-18	19-25	26-55	56-64	65+
1	6	14	75	15	16
2	4	5	44	18	7
3	4	5	53	21	15

Table 4. 54 Frequency (N) of the variation type of the confluence of sinuses for age in ten-year intervals.

Variation type	Age in ten-year intervals					
	0-19	20-29	30-39	40-49	50-59	60+
1	9	17	20	24	31	25
2	6	9	18	10	23	12
3	6	11	16	23	12	30

There is no significant association of variations within the confluence of sinuses and the presence of pathology (p=0.236), see table 4.55 below.

Table 4. 55 Frequency (N) of the variation type of the confluence of sinuses and presence of pathology.

Variation type	Pathology	
	Present	Absent
1	97	23
2	64	12
3	84	10

There is no significant association of variations within the confluence of sinuses and the type of circulatory dominance (p=0.140), see table 4.56 below.

Table 4. 56 Frequency (N) of the variation type of the confluence of sinuses and dominance.

Variation type	Dominance		
	Left dominant	Co-dominant	Right dominant
1	23	50	53
2	19	35	24
3	15	34	49

<u>**4.9 Other variations of the dural venous sinuses**</u>

As described in section 3.2.3 above a list of 20 types of possible variations was created for the study, however, not all 20 types were identified in the current study.

Table 4. 57 Frequency (N) of variation types in the current sample.

Type	Variation	Frequency (N) in current study	Reference in the literature
1	Absence of the transverse sinus on the right	2 (0.7%)	- 3 out of 50 cases (6%) (Widjaja & Griffiths, 2014) - 3 out of 211 cases (1.4%) (Rollins *et al.*, 2005) - 10 out of 100 cases (10%) (Bayaroğullari *et al.*, 2018) - 4 out of 105 cases (3.8%) (Alper *et al.*, 2004) - 12 out of 1654 (0.7%) (Goyal *et al.*, 2014)
2	Absence of the transverse sinus on the left	3 (0.9%)	- 26 out of 100 cases (26%) (Bayaroğullari *et al.*, 2018) - 19 cases out of 105 (18.1%) (Alper *et al.*, 2004) - 67 out of 1654 (4.1%) (Goyal *et al.*, 2014)
3	Absence of the transverse sinuses bilaterally	0	-1 out of 50 cases (2%) (Widjaja &Griffiths, 2014)
4	Multiple transverse sinuses unilaterally	0	- 1 case report of a fenestrated transverse sinus (Massrey *et al.*, 2018)
5	Multiple transverse sinuses bilaterally	0	- None reported in the reviewed literature
6	Right dominant transverse sinus	126 (41.7%)	- 36 out of 100 cases (36%) (Kitamura *et al.*, 2017) - 73 out of 100 cases (73%) (Sharma & Sharma, 2012) - 62 out of 102 cases (60.8%) (Manara *et al.*, 2010) - 11 out of 31 cases (35.5%) (Park *et al.*, 2008) - 59 out of 100 cases (59%) (Surendrababu *et al.*, 2006) - 59 out of 100 cases (59%) (Ayanzen *et al.*, 2000) - 22 out of 189 cases (11.6%) (Durgan *et al.*, 1993) - 51 out of 100 cases (51%) (Browning, 1953) - 13 out of 25 cases (52%) (Gibbs *et al.*, 1934) - 12 out of 25 cases (48%) (Edwards, 1931)

Type	Variation	Frequency (N) in current study	Reference in the literature
7	Left dominant transverse sinus	57 (18.9%)	- 14 out of 100 (14%) cases (Kitamura *et al.*, 2017) - 18 out of 100 cases (18%) (Sharma & Sharma, 2012) - 17 cases out of 100 (17%) (Manara *et al.*, 2010) - 30 out of 100 cases (30%) (Surendrababu *et al.*, 2006) - 25 out of 100 (25%) (Ayanzen *et al.*, 2000) - 10 out of 189 cases (5.3%) (Ayanzen *et al.*, 2000) - 29 out of 100 cases (29%) (Browning, 1953) - 6 out of 100 cases (6%) (Gibbs *et al.*, 1934) - 11 out of 25 cases (44%) (Edwards, 1931)
8	Codominance of the transverse sinuses	119 (39.4%)	- 50 out of 100 cases (50%) (Kitamura *et al.*, 2017) - 1106 out of 1654 cases (66.9%) (Goyal *et al.*, 2016) - 9 out of 100 cases (9%) (Sharma & Sharma, 2012) - 18 out of 31 cases (58.1%) (Park *et al.*, 2008) - 10 out of 100 cases (10%) (Surendrababu *et al.*, 2006) - 33 out of 105 cases (31.4%) (Alper *et al.*, 2004) - 16 out of 100 cases (16%) (Ayanzen *et al.*, 2000) - 20 out of 189 cases (10.6%) (Ayanzen *et al.*, 2000) - 20 out of 100 cases (20%) (Browning, 1953) - 6 out of 25 (24%) cases (Gibbs *et al.*, 1934) - 2 out 25 cases (8%) (Edwards, 1931)
9	Multiple SSS	0	- None reported in the reviewed literature
10	Bifurcation of the SSS into left and right transverse sinuses (no confluence) – corresponds to SSS drainage patterns type 4,7 and 8	64 (21.2%)	- 10 out of 100 cases (10%) (Bisaria, 1985) - 1 case reported by Özen *et al.*, 2013 - 1 case reported by Hosley *et al.*, 1991

Type	Variation	Frequency (N) in current study	Reference in the literature
11	Bifurcation of the SS into left and right transverse sinuses – corresponds to SSS drainage patterns type 2 and 3	112 (37.1%)	- 10 out of 100 cases (10%) (Bisaria, 1985)
12	SSS draining into only one of the transverse sinuses – corresponds to SSS drainage pattern types 2,3 and 9	113 (37.4 %)	- Bifurcation of the SSS may be seen as duplication in the region of the confluence of sinuses (Tubbs *et al.*, 2019), full duplication of the SSS not reported in the reviewed literature
13	SSS draining into only one of the transverse sinuses and the SS draining into the other – corresponds to SSS drainage pattern type 5 and 6	13 (4.3%)	- 20 out of 100 cases (20%) the SSS drained to the right transverse and the SS drains to the left transverse sinus (Bisaria, 1985) - 7 out of 100 cases (7%) the SSS drains to the left transverse and the SS drains to the right transvers sinus (Bisaria, 1985)
14	Absence of sigmoid sinus on the right	0	- 7 out of 211 cases (3.31%) (Bayaroğullari *et al.*, 2018)
15	Absence of the sigmoid sinus on the left	2 (0.7%)	- 2 out of 211 cases (0.94%) (Bayaroğullari *et al.*, 2018) - Absent unilaterally in 236 out of 1654 (14.3%) cases (Goyal *et al.*, 2014)
16	Absence of sigmoid sinus bilaterally	0	- None reported in the reviewed literature
17	Absent inferior sagittal sinus	2 (0.7%)	- 89 out of 100 cases (89%) (Sharma & Sharma, 2012) - 48 out of 100 cases (48%) (Ayanzen *et al.*, 2000) - 120 out of 211 cases (56.9%) (Bayaroğullari *et al.*, 2018)
18	Absent SS	0	- 9 out of 100 cases (9%) (Sharma & Sharma, 2012)
19	Multiple SS	0	- 3 out of 100 cases (3%) (Bisaria, 1985)
20	Variations in the configuration of the confluence of	78 (25.8%)	- 235 out of 394 cases (59.6%) (Gökçe *et al.*, 2014)

| sinuses, including partitioning into compartments | | - 161 out of 211 cases (76.3%) (Bayaroğullari *et al.*, 2018) |

4.10 Venous blood pools

After careful examination of the tentorium cerebelli, ventricles and venous sinuses no structures which could be referred to as a venous blood pool was found.

4.11 Presence of pathology

In 290 cases the presence or absence of a pathology was known. Of these cases, the majority (245; 84.48%) exhibited some form of pathology, refer to graph 4.6 below.

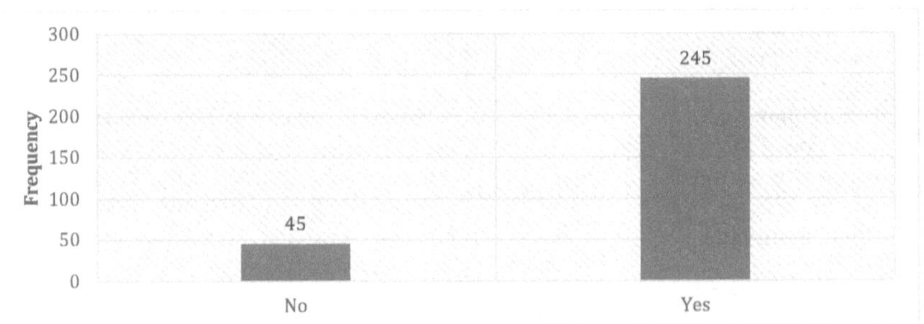

Graph 4. 6 Graph showing the presence of pathology within the sample.

There is no significant difference in the median age between participants displaying pathology and those not presenting with pathology (p=0.052).

There is no association between the presence of pathology and age in terms of stage of life (p=0.199; table 4.58) or age in ten-year intervals (p=0.149; table 4.59).

Table 4. 58 Table showing the presence of pathology across ages (stage of life).

	0-18	19-25	26-55	56-64	65+	Total
Pathology Present	11	18	134	48	34	245
Pathology Absent	2	5	31	4	3	45

Table 4. 59 Table showing the presence of pathology across ages (ten-year interval).

	0-19	20-29	30-39	40-49	50-59	60+	Total
Pathology Present	17	27	45	40	57	59	245
Pathology Absent	3	8	8	13	8	5	45

A significant association exists between sex and the presence of pathology (p=0.02), refer to table 4.60 below.

Table 4. 60 Frequency (N) of pathology in males and females.

	Sex	
Pathology	**Female**	**Male**
Present	140	11
Absent	34	105

Logistic regression revealed that males are 2.3 (95% CI: 1.12 ; 4.79) times more likely to present with pathology than females, refer to table 4.61 below.

Table 4. 61 Logistic regression showing males are 2.3 times more likely to present with pathology than females (copied from statistical analyses report for the current study).

```
Logistic regression                          Number of obs    =        290
                                             LR chi2(1)       =       5.67
                                             Prob > chi2      =     0.0173
Log likelihood =  -122.3232                  Pseudo R2        =     0.0226


------------------------------------------------------------------------------
       path | Odds Ratio   Std. Err.      z    P>|z|     [95% Conf. Interval]
------------+-----------------------------------------------------------------
      1.Sex |   2.318182    .8580014     2.27   0.023     1.122283    4.788425
      _cons |   4.117647    .7872638     7.40   0.000     2.830777    5.989528
------------------------------------------------------------------------------
```

There is no significant association between circulatory dominance and the presence of pathology (p=0.998).

There is no significant association between pathology and any variations of the SSS sinus (p=0.491) or within the confluence of sinuses (p=0.236).

4.12 Adjusted vessel volumes

Vessel volumes for males were reduced in size by 10% and all vessel volumes re-analysed to determine if the difference in size between males and females may be the underlying reason for the differences seen between subjects when analysed by sex (table 4.62 below).

Table 4. 62 Table showing adjusted (decreased) median vessel volumes for males versus the unadjusted vessel volumes.

	Median Vessel Volumes				
Sex	**R Transverse**	**L Transverse**	**R Sigmoid**	**L Sigmoid**	**SSS**
Unadjusted volumes					
Male	2.138	1.6865	2.8775	2.11225	5.5012
Female	1.559	1.192	2.4583	1.9315	4.4192
% difference	-27.08%	-29.32%	-14.57%	-8.56%	-19.67%
Adjusted volumes					
Male	1.9242	1.517985	2.58975	1.901025	4.95108
Female	1.559	1.192	2.4583	1.9315	4.4192
% difference	-18.98%	-21.47%	-5.08%	1.6%	-10.74%

This analysis altered the significance for the right sigmoid sinus and not the other sinuses. The differences between sexes were significant for the transverse and SSS and not significant for the sigmoid sinuses, see table 4.63 below. Even with the 10% adjustment the males had greater vessel volumes than the females.

Table 4. 63 Table showing the significant differences between male and female median vessel volumes.

Vessel	Median volume (cm³)		p-value
	Male	Female	
Transverse (R)	1.9242	1.559	0.0004
Transverse (L)	1.517985	1.192	0.0188
Sigmoid (R)	2.58975	2.458	0.5060
Sigmoid (L)	1.901025	1.932	0.7592
Superior Sagittal Sinus	4.95108	4.419	0.008

A matched pair analysis (Wilcoxon sign rank test) was used to determine if a significant difference existed between the median volume of vessels on the right side and vessels on the left side within individuals. Significant differences were noted in the median volume of the left and right sigmoid sinus (p<0.001; table 4.64) as well as the median volume of the left and right transverse sinus (p<0.001; table 4.65).

Table 4. 64 Significant difference in the vessel volume between the right and left sigmoid sinuses (copied from statistical analyses report for the current study).

```
Paired t test
------------------------------------------------------------------------------
Variable |     Obs        Mean    Std. Err.    Std. Dev.   [95% Conf. Interval]
---------+--------------------------------------------------------------------
 Vol~Radj |     292    2.562906    .0554073    .9468008    2.453856    2.671956
 Vol~Ladj |     292    2.046483    .0508222    .8684498    1.946458    2.146509
---------+--------------------------------------------------------------------
    diff |     292    .5164226    .0814637    1.392053    .3560898    .6767554
------------------------------------------------------------------------------
    mean(diff) = mean(VolSigRadj - VolSigLadj)                 t =   6.3393
Ho: mean(diff) = 0                              degrees of freedom =      291

Ha: mean(diff) < 0             Ha: mean(diff) != 0              Ha: mean(diff) > 0
Pr(T < t) = 1.0000          Pr(|T| > |t|) = 0.0000           Pr(T > t) = 0.0000
```

Table 4. 65 Significant differences in the vessel volume between the right and left transverse sinuses (copied from statistical analyses report for the current study).

```
Paired t test
------------------------------------------------------------------------------
Variable |     Obs        Mean    Std. Err.    Std. Dev.   [95% Conf. Interval]
---------+--------------------------------------------------------------------
VolRtadj |     295    1.813256    .0491195     .8436556    1.716586    1.909926
VolLtadj |     295    1.484637    .0483067     .8296944    1.389566    1.579707
---------+--------------------------------------------------------------------
    diff |     295    .3286194    .0607042    1.042629     .2091495     .4480892
------------------------------------------------------------------------------
     mean(diff) = mean(VolRtadj - VolLtadj)                      t =     5.4135
Ho: mean(diff) = 0                              degrees of freedom =        294

Ha: mean(diff) < 0            Ha: mean(diff) != 0               Ha: mean(diff) > 0
Pr(T < t) = 1.0000         Pr(|T| > |t|) = 0.0000            Pr(T > t) = 0.0000
```

A Kruskal Wallis test indicated significant difference in the median volume of the SSS between age groups (p= 0.0038). A Spearman's rank correlation indicated no association with vessel volume and age for all vessels except the SSS. A weak negative association (r_s = -0.1762, p = 0.0022) was noted between the volume of the SSS and age.

No significant differences were noted in the median volume of vessels between patients displaying pathology and those not displaying pathology.

For the presence of accessory vessels, significant differences were only seen in the median volume of the right transverse sinus (p=0.0033) and SSS (p=0.0117) when the left vein of Trolard was present or absent. In both cases the volumes were significantly lower in individuals where the vein was absent. These results should be interpreted with caution due to the low sample size of cases where the left vein of Trolard was absent.

Variation in the confluence of sinuses had a significant effect on the volume of the right sigmoid sinus and not on any of the other sinuses. A significant difference was noted in the median volume of the right sigmoid sinus (p=0.0257). Post hoc pairwise tests indicated that individuals with variation type 3 had a significantly increased volume in the sigmoid sinus compared to those with no variation (p=0.041).

A significant difference was seen in the median volume of vessels between the different types of variations of the SSS. Differences were noted in the volume of the right transverse sinus (p=0.0001), left transverse sinus (p=0.0001), right sigmoid sinus (p=0.0001), and the left sigmoid sinus (p=0.0001) between the variations of the SSS.

Pairwise differences can be seen in the table 4.66 below (only statistically significant differences displayed). Important to note that the sample size for variation types 5 – 9 were small, therefore even large differences in these vessels are unlikely to be significant.

Table 4. 66 Significant difference in the median vessel volume between the different types of variations of the SSS.

Vessel and SSS variation type	Median volume (cm3)	p-value*
Transverse (R) *No variation* *Type 3* *Type 4*	2.3885 1.8440 2.2987	0.002 0.001
Transverse (L) *No variation* *Type 2*	1.4693 0.8965	<0.001
Sigmoid (R) *No variation* *Type 2*	2.3885 3.9687	<0.001
Sigmoid (L) *No variation* *Type 2* *Type 3*	2.0038 1.6089 2.4995	0.007 0.035

* pairwise comparison to median vessel volume with no variation in the SSS present. Bonferonni correction applied

Significant differences were noted in the median volume of vessels and the presence of hypoplasia on the right and left sides (see table 4.67 below). Typically, the presence of hypoplasia on the left side was associated with a lower volume of the left transverse and sigmoid sinuses and a greater volume in the right sigmoid sinus. The presence of hypoplasia on the right side was associated with a lower volume of the right transverse and sigmoid sinuses and a greater volume in the left sigmoid sinus.

Table 4. 67 Significant differences in the median vessel volume and the presence of hypoplasia.

Vessel volume	Median volume (cm³)		p-value
	Hypoplasia Absent	**Hypoplasia Present**	
Left side hypoplasia			
Transverse (R)	1.58	1.73	0.6014
Transverse (L)	2.64	1.19	<0.0001
Sigmoid (R)	1.84	2.51	<0.0001
Sigmoid (L)	2.49	1.86	0.0003
Superior Sagittal Sinus	4.33	4.75	0.4389
Right side hypoplasia			
Transverse (R)	2.67	1.53	<0.0001
Transverse (L)	1.14	1.37	0.10
Sigmoid (R)	3.11	2.35	<0.0001
Sigmoid (L)	1.53	2.07	<0.0001
Superior Sagittal Sinus	4.39	4.76	0.0308

The effect of changing the male volumes overall is summarised in table 4.68 below.

Table 4. 68 Summary of the differences between unadjusted male vessel volumes and the 10% adjusted male vessel volumes.

Variable	Unadjusted volumes	Adjusted volumes	Difference
Median vessel volume between the sexes	Significant difference for the median vessel volumes of the right and left transverse, right sigmoid and SSS.	Significant difference for the median vessel volume of the right and left transverse and SSS.	Median volume of the right sigmoid sinus was significantly different between males and females for the unadjusted sample and not significantly different for the adjusted sample.
Vessel volume and sex	Vessels in males are shown to have greater volumes than the vessels of females.	Males had greater vessel volumes than the females.	No difference.
Vessel volume between right and left sides	Significant differences were seen for the volume of the right versus left sigmoid sinus as well as the median volume of the right versus left transverse sinus.	Significant differences were noted in the median volume of the left and right sigmoid sinus as well as the median volume of the left and right transverse sinus.	No difference.
Vessel volume and age	A significant difference in the median volume of the SSS between age groups was seen.	A significant difference was seen in the median volume of the SSS between age groups.	No difference.
Vessel volume and pathology	No association was seen between the volume of the sinus and the presence or absence of pathology in the sample.	No association between vessel volume and the presence of pathology.	No difference.
Vessel volume and presence of accessory vessels	A significant difference in the median volume of the right transverse sinus and the SSS was seen when the left vein of Trolard was either present or absent.	Significant differences were seen in the median volume of the right transverse sinus and SSS when the left vein of Trolard was present or absent.	No difference.
Vessel volume and variation in the confluence sinuses	Variation in the confluence of sinuses had a significant effect on only the right and left sigmoid sinuses and not on the other sinuses.	Variation in the confluence of sinuses did not have a significant effect on the volume of any vessels except for the right sigmoid sinus.	No difference.

Variable	Unadjusted volumes	Adjusted volumes	Difference
Vessel volume and variation in the drainage patter of the SSS	Significant differences in the median volumes of the right and left transverse sinuses, , right and left sigmoid sinuses were seen between the different variation types of the SSS.	Differences were noted in the volume of the right and left transverse sinuses, right and left sigmoid sinuses between the variations of the SSS.	No difference.
Vessel volume and hypoplasia	Significant differences were noted in the median volume of vessels and the presence of hypoplasia on the right and left sides.	Significant differences were seen in the median volume of vessels and the presence of hypoplasia on the right and left sides.	No difference.

4.13 Venous sinus configuration as a method of identification

Each patient was assigned a 6-digit identification number. However, it was found that the numbers repeated with some patients having the same I-Number. Persons with more common venous drainage patterns shared various numbers within the sequence and thus an entire sequence could be shared between individuals.

There were instances where the I-Number was repeated in the sample of 302 patients. For the total sample site 195 shared their I-Number with at least one other person of which the 33 of repetitions occurred between one and five times and 9 of the repetitions occurred between six and nine times. Only six I-Numbers repeated 10 or more times, refer to table 4.69 below.

A complete list of I-Numbers is available in appendix C.

Table 4. 69 Table showing the frequency (N) of the I-Numbers which repeated the most in the current study.

I-Number	Description	Frequency (N)
231230	Female, left hypoplastic, right dominant, SSS drains to RT and SS drains to both RT and LT, non-confluence, no other variation	10
113110	Male, bilateral hypoplasia, co-dominant, general SSS drainage pattern, true confluence, no other variation	11
213420	Female, bilateral hypoplasia, co-dominant, bifurcation of SSS to the RT and LT, partial confluence, no other variation	13
211110	Female, bilateral hypoplasia, right dominant, general SSS drainage pattern, true confluence, no other variation	14
211230	Female, bilateral hypoplasia, right dominant, SSS drains to the RT and straight drains to RT and LT, non-confluence, no other variation	19
213110	Female, bilateral hypoplasia, co-dominant, general SSS drainage pattern, true confluence, no other variation	22

Chapter 5: Discussion

5.1 Fetal dissection

A total of five stillborn fetuses from the New Somerset Hospital in Cape Town were dissected. The purpose of the dissections was to compare a local sample to currently available literature in order to give a complete view of the venous drainage system from birth until adulthood. The fetal dissections were not included in the statistical analyses. The dissection aided the researcher to become familiar with the gross structure of the venous sinuses of fetuses.

The superior sagittal, transverse, straight and confluence of sinuses could clearly be seen once the cerebrum was removed. For all fetuses a type 1 confluence was seen (the generally acceptable configuration), in which the SSS drained directly into the confluence and did not bifurcate into the transverse sinuses. It is interesting to note that none of the fetuses had a bifurcating SSS or an SSS that drained unilaterally, as was seen in the adult sample.

Blood could be seen between the layers of the tentorium cerebelli. These blood pools are thought to be the remnants of embryological development; refer to section 5.8.1 for further discussion.

5.2 Epidemiological variables

5.2.1 Sex

There were more females (181) than males (121) in the sample. This is thought to be due to the propensity of females in seeking health care services compared with males (Verbrugge *et al.*, 1987; Hunt *et al.*, 1999).

5.2.2 Age

The minimum age was 13 years and the maximum age was 81 years, with a mean age of 45.39 years. The high mean age for the sample was expected as MRIs, especially with contrast, is very seldom done in children. This is due to long exposure times, children having to be anesthetised in order to obtain a quality image and also due to the cost involved.

The median age of females (50 years) in the study is significantly greater than the median age of males (43 years) (p= 0.0143). It is thought that males presented earlier to hospital with pathology than females due to lifestyle differences, as men are more likely to smoke and use alcohol and be physically less active than females (Fiala & Brázdová, 2000; Vari *et al.*, 2016) It is also interesting to note that the median age of females corresponds to when females enter menopause and a decrease in the level of oestrogen (which has certain protective properties) may be why females present to hospital at a later age than males (Mendelsohn & Karas, 1999; Iorga *et al.*, 2017).

5.3 Angiographic variables

5.3.1 Venous sinus volume

Tracing the venous sinus volume and using it for statistical and morphological analyses is a novel way of looking at the dural venous sinuses. The study was the first to use venous volume of a dural sinus in describing variations and dominance of the dural venous sinuses.

The superior sagital sinus was found to have the greatest maximum volume (10.8275 cm^3), followed by the right sigmoid (6.8563 cm^3) and left transverse sinuses (6.5815 cm^3). Although the left transverse sinus had the third largest maximum volume (after SSS and R. sig) it had a smaller mean volume (1.5537 cm^3) compared to the left sigmoid sinus (2.1451 cm^3). One would expect the sigmoid sinuses to have larger mean volumes than the transverse sinuses, as the sigmoid sinuses receives blood from the transverse, inferior and superior petrosal sinuses. This is further demonstrated by the mean vessel volumes of the sigmoid sinuses, where the mean vessel volumes were found to be larger than that of the transverse sinuses.

An investigation was carried out to compare vessel volume between the right and left sides for each individual. A significant difference was seen between the right and left sigmoid sinuses and between the right and left transverse sinuses. In both cases, the sinus on the right was significantly larger than the sinus on the left. This corresponds with the finding that the majority of persons in the sample were found to have right sided dominance (highest volume of blood drains through the sinuses on the right), refer to section 5.4 below. Furthermore, the right transverse sinus had a larger mean vessel volume when compared to the left transverse sinus, and this confirms the finding that most persons in the current sample were right side dominant.

5.3.2 Venous sinus volume and sex

The human brain at birth is 25% of its adult weight in adult males (1336 grams) and females (1198 grams) respectively according to Hartmann *et al.,* 1994.

In this study males were found to have significantly greater sinus volumes than females. An exception to this was the left sigmoid sinus, even though the males had a larger median volume (2.112 cm^3) compared to females (1.932 cm^3) the difference was not statistically significant. The difference in vessels volume could be due to males having a larger brain volume (Dekaban, 1978).

The volumes of the vessels in males were decreased by 10% to establish if general size of a person contributed to the difference in vessel volume between the sexes. A significant difference was still found between males and females for the transverse and SSS. However, the difference between males and females for the right sigmoid sinus was no longer found to be significant. It is speculated that there might be another underlying reason why males exhibit significantly larger vessel volumes than females. This underlying reason might be a hormonal influence, as males have higher testosterone levels than females leading to males being larger in body size and having a heavier brain than females.

5.3.3 Venous sinus volume and age

For the comparison between vessel volume and age the SSS volume of one patient could not be used due to the sinus being obscured on image and the vessel volume could not be traced. Thus only 301 subjects' data were used for SSS volume compared to age.

No significant association was found between any of the vessel volumes and age, except for the SSS. There was no significant association between vessel volume and age for the transverse and sigmoid sinuses. However, there was a significant association for the SSS for which a weak negative association was found.

A significant difference in the median volume of the SSS between the age groups was found (for stage of life and ten-year intervals). The SSS was shown to decrease in volume as the subgroups increased in age.

As this is not a longitudinal study, one cannot plot the changes in the SSS over time. However, a change in structure and function is seen in the villi as we age. Mineralisation of the villi leads to the formation of granules which in turn causes a reduced reabsorption of CSF (Gomez *et al.*, 1983; Pollay, 2010). As the volume in the SSS is composed of blood and CSF; it might be that with age less CSF is reabsorbed into the SSS and therefore we see a significant decrease in volume in older persons. In older people there is a greater reliance on other systems for the re-absorption of CSF, for example the up regulation of the g-lymphatic system (Jessen *et al.*, 2015).

5.3.4 Venous sinus volume and pathology

Pathology was not seen to influence the median volumes of the cerebral venous sinuses for the sample, also, no association was found between the volume of the sinus and the presence or absence of pathology in the individual. It is hypothesised that the venous drainage pattern has a larger influence on the vascular interventions in the treatment of pathology than the blood volume of the vessels.

5.3.5 Venous sinus volume and the presence of accessory vessels

The median volumes of the right transverse and the SSS were significantly lower when the left vein of Trolard was absent. The veins of Trolard, Labbé and Sylvian (superficial middle cererbal vein) are interconnected in the region of the posterior lateral sulcus. They form a balanced pattern of venous drainage. However, in some cases one of the veins may be larger (more dominant) than the others (Shapiro, 2020). It is thought that in the current sample the veins of Trolard were dominant, especially the left vein of Trolard, and thus when it was absent it caused a significant decrease in the blood flow to the SSS and in turn the right transverse sinus.

The presence or absence of the veins of Labbé did not have a significant influence on the volumes of the venous sinuses. This could be due to the vein of Labbé opening into the transverse sinus and thus having no effect on the blood draining from the SSS towards the transverse and in turn the sigmoid sinuses.

The internal cerebral veins drain into the vein of Galen which in turn drains to the SS. As these veins were present in almost all cases, we could not speculate as to what would happen if they were absent in a significant number of patients.

5.3.6 Venous volume and variation in the confluence of sinuses

Variation in the confluence of sinuses was clasified into three types:

Type 1: True confluence
Type 2 :Ring structure present within the confluence
Type 3: No confluence

The majority of patients had a true confluence of sinuses (126), followed by patients who had no confluence (98) and patients with a ring structure in the confluence (78), refer to table 5.1 below.

Table 5. 1 Frequency of the different types of the configuration of the confluence of sinuses expressed as a percentage

Configuration of confluence of sinuses	Current study	Gökçe *et al.,* (2014)	Bayaroğullari *et al.,* (2018)
Type 1	41.7%	37.8%	26.1%
Type 2	32.5%	59.6%	59.7%
Type 3	25.8%	2.5%	14.2%

The current study found the type 1 configuration to be the most common, whereas Gökçe *et al.,* (2014) and Bayaroğullari *et al.,* (2018) found the type 2 to be the most prevalent in their samples. The studies by Gökçe *et al.,* (2014) and Bayaroğullari *et al.,* (2018) where both conducted in Turkey, the difference may be due to the different populations studied, and that the type 1 configuration is more prevalent in a South African sample.

Statistical analyses showed that persons with no confluence had a significant increase in the median volume of the right sigmoid sinus compared to those without variation in the confluence (type 1). The absence of a confluence is usually accompanied by the SSS only draining into the right transverse sinus (58 out of 98 cases); which in turn leads to an increase of blood flow to the right sigmoid sinus. However, in cases where a type 1 configuration of the SSS was seen (draining directly into the confluence), a significant increase in the volume of the right transverse sinus was not seen, suggesting a more equal distribution of blood flow to the two transverse sinuses.

5.3.7 Venous sinus volume and variations in the drainage pattern of the superior sagittal sinus

Variations in the drainage pattern of the SSS did not significantly influence the volumes of the SSS. It is thought that the volume of blood flowing through the SSS and into the transverse sinuses is not influenced by variations in the drainage pattern of the SSS, as the amount of blood draining from the SSS will still drain to the transverse sinuses irrespective of the configuration of the SSS.

The current study compared the volumes of the other venous sinuses with variations of the SSS, to determine whether variations of the drainage pattern of the SSS influenced the volumes of the other venous sinuses. A significant difference was seen for the volume of all the other sinuses and thus variations in the drainage pattern of the SSS influenced the volume of blood throughout the other sinuses. This is expected because the SSS is the major contributor of blood flow to the transverse and sigmoid sinuses.

A significant association was seen between the median volume of the right transverse sinus and variation types 3 and 4 of the SSS. Variation type 3 (34 cases) determines that the SSS drains only to the left transverse sinus, thus one would expect a significant decrease in the blood volume through the right transverse sinus, as is seen with a p value of 0.002. Another significant association was also seen between the volume of the right transverse sinus and variation type 4 (54 cases) of the SSS, in these cases the SSS bifurcates and drains to the right and left transverse sinuses respectively, without entering the confluence of sinuses. This bifurcation of the SSS may lead to blood being distributed unevenly between the transverse sinuses, thus leading to right (13 out of 54 cases) or left (15 out of 54 cases) dominance; when the blood flow is distributed more evenly one would expect to find a co-dominant (26 out of 54 cases) venous flow.

There was a significant difference in the volume of the left transverse sinus when a type 2 variation (78 cases) of the SSS was present; in these cases the SSS only drained into the right transverse sinus, while the SS drained into both the left and right transverse sinuses. Fifty-eight of the 70 cases with a type 2 variation of the SSS were found to be right dominant, only two cases were left dominant and 18 cases were co-dominant. In both cases where left dominance was seen, the patients' left transverse and sigmoid sinuses were much larger than the transverse and sigmoid sinuses on the right, which could explain why a left sided dominance is seen in these persons and not right sided dominance. One could argue that in exceptional cases the blood flow from the SSS alone does not determine dominance in an individual.

A type 2 variation of the SSS had a significant influence on the median volume of the right sigmoid sinus. As the SSS drains only to the right transverse (78 cases) one whould expect an increase in volume into the right sigmoid sinus.

The left sigmoid sinus showed a significant difference in its median volume when associated with a type 2 or 3 variation on the SSS. With a type 2 variation the SSS only drains to the right transverse sinus, thus less blood drains through the left transverse and consequently less blood flows through the left sigmoid sinus. On the other hand a type 3 variation of the SSS, has the SSS only draining to the left transverse sinus (34 cases), which in turn results in more blood flow through the left sigmoid sinus, and one whould expect an increase in the median volume of the left sigmoid sinus. Of the 34 cases where the SSS drained only to the left transverse sinus, 15 were left dominant, one was right dominant and 18 were co-dominant.

The sample sizes for variations type 5 – 9 were small, therefore even if there were large differences in the volumes of the sinuse it did not show statistical significance on analyses. Only one case was identified with a type 9 drainage pattern. In this case both the SSS and SS drained to the right transverse sinus. This differs from the finding of Browning in 1953 who identified three such cases in all the SSS and SS drained to the left transverse sinus. This difference could be due to the difference in the populations studied.

The combined volumes of the transverse and sigmoid sinuses on the left and right sides, showed a significant difference in the median volume of the SSS in persons with a type 2 or 3 variation. In type 2 and 3 variations of the SSS, the SSS drains only to one side. Thus, it is predicted that the volume of the transverse sinus and ultimately the sigmoid sinus will significantly different between persons with a type 2 or 3 variations and those without.

When the pooled right and pooled left volumes were compared between the sexes, males were shown to have significantly larger vessel volumes than females. This could be due to the difference between male and female morphology and males having generally larger blood volumes. It has also been shown that there is a difference in overall body size and weight of the brain when comparing males and females.

5.4 Dominance and the cerebral venous system

The asymmetry of the SSS is found in the fetus as well as in the adult; this is seen as a tendency to drain more to one side of the head than to the other. This pattern already seen in the fetus by the time it is 20 mm in length; this is predominantly to the right side. (Streeter, 1915).

For the current study dominance was expressed as a range rather than a set value, refer to graph 5.1 below.

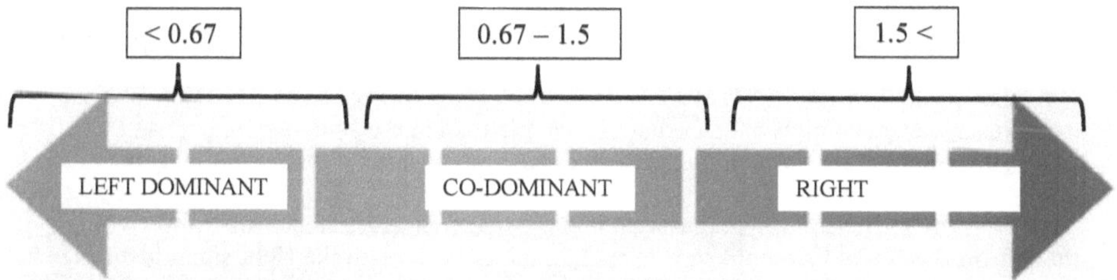

Graph 5. 1 Dominance expressed on a continuous scale.

The majority of individuals displayed right-sided dominance (41.72%) followed by co-dominance (39.40%) and left-sided dominance (18.87%). These three values fall within the ranges found by other authors. This is also in line with literature which states that right-sided dominance is most prevalent as well as the accepted anatomy that the SSS drains pre-dominantly into the right transverse sinus. In the literature right-sided dominance was identified in a range between 11.6% (Durgan *et al.*, 1993) and 73% (Sharma & Sharma, 2012). Left-sided dominance was identified at between 5.3% (Ayanzen *et al.*, 2000) and 44% (Edwards, 1931). Co-dominance was identified between 31.4% (Alper *et al.*, 2004) and 66.9% of cases (Goyal *et al.*, 2016). For a detailed breakdown of findings by other authors refer to table 4.57 above.

According to the studies by Kitamura *et al.*, (2017) and Manar *et al.*, 2010 there were no statistically significant associations between age or sex and venous sinus dominance. Similar to the literature, the current study also revealed no statistically significant association between circulatory dominance and age or sex. It is thought that the drainage pattern of the SSS into the transverse sinuses, and thus dominance, is already established at birth and thus that age and sex would not alter this drainage pattern.

No significant association was found between the type of circulatory dominance and the presence of pathology; for further discussion on pathology refer to section 5.8.

The presence of accessory vessels was shown to have an influence on the median volume of the venous sinuses. It is thought that if an accessory vessel is absent less blood would then drain to the corresponding venous sinus. However, the presence of an accessory vessel did not influence the dominance of an individual as it is thought that the contribution to volume which is made by an accessory vessel is not significant enough to alter the dominance of that individual.

5.5 Hypoplasia and aplasia

5.5.1 Hypoplasia of the right and left transverse sinuses

In the current study 296 individuals had data available for the left transverse sinus and 299 individuals had data available for the right transverse sinus. Of the individuals for which data was present, 293 displayed a form of hypoplasia. Of these individuals, the majority (207) had hypoplasia of both the left and right transverse sinuses. Statistical analyses showed a significant association between hypoplasia on the left and right side (p=0.0004). Thus, a significant association exists between having hypoplasia on the one side as well as on the other. It is expected that the factors which influence hypoplasia on the one side will also have an impact on the vessels of the other side. In the study by Widjaja and Griffith (2004) they only examined 50 patients and found that in 26 patients (52%) one of the transverse sinuses was hypoplastic. In the current study 78% of patients (234/299) showed hypoplasia on the right and 89% (266/296) showed hypoplasia on the left. In the current study most individuals displayed hypoplasia with 99% (293/294) of subjects showing some form of hypoplasia. The presence of hypoplasia was much higher for the current sample than in the study by Widjaja and Griffiths (2004) done in England, possibly suggesting or geographical difference in the study population.

Typically, the presence of hypoplasia on the left side was associated with a lower volume of the left transverse and sigmoid sinuses and a greater volume in the right sigmoid sinus. The presence of hypoplasia on the right side was associated with a lower volume of the right transverse and sigmoid sinuses and a greater volume in the left sigmoid sinus. A hypoplastic vessels limits the volume of blood flow through the vessel, thus causing increased blood flow to the contralateral side.

5.5.2 Hypoplasia and sex

A significant association was found between the sex of an individual and hypoplasia on the right side. Significantly more females displayed hypoplasia on the right than compared to males. No significant association was seen between sex and hypoplasia on the left. The difference between males and females could be due to hormonal influence or genetic influence. The exact extend to which genetics and hormones influence venous sinus development would require further investigation.

5.5.3 Hypoplasia and age

No association was found between age and hypoplasia. It is thought that one is born with hypoplastic vessels, and it is not something that develops over time. In the current study older people did not have significantly more hypoplastic vessels than the younger persons, and this suggests that the presence of hypoplastic vessels does not change over time.

5.5.4 Hypoplasia and circulatory dominance

A clear association was seen between hypoplasia and the type of circulatory dominance. Left side hypoplasia is significantly more common in persons with right side circulatory dominance. Right side hypoplasia is significantly more common in persons with left side circulatory dominance. A logical association can thus be made between circulatory dominance on the one side and hypoplasia on the other. A person who is right side dominant is likely to have hypoplasia of the venous sinuses on the left.

5.5.5 Hypoplasia and pathology

The presence of one or more hypoplastic vessels did not show an association with the presence with pathology.

5.5.6 Hypoplasia of the transverse sinuses and presence of accessory vessels

No association was found between the presence of accessory vessels and hypoplasia of the transverse sinuses. For this study hypoplasia of the transverse sinuses were defined as occurring when the volume of the any one of the transverse sinuses were less than half the volume of the SSS. As only the veins of Labbé drain directly into the transverse sinuses they could therefore potentially have a direct effect on the volume of the transverse sinuses and hypoplasia. However, as no association was found between the presence of the veins of Labbé and hypoplasia it is thought that the volume contribution that these veins make are too small to cause the transverse sinus to become hypoplastic.

5.5.7 Hypoplasia and variation of the confluence of sinuses

No association was found between variations in the pattern of formation of the confluence of sinuses and the presence of hypoplasia. Thus, none of the variations of the confluence contributed significantly to a transverse sinus being hypoplastic. The confluence may be seen as a distribution point of the blood flow from the SSS to the transverse sinuses, thus variation of the confluence does not influence the volume of blood draining from the SSS to the transverse sinuses.

5.5.8. Hypoplasia and variation of the superior sagittal sinus

A significant association was found between the drainage pattern of the SSS and hypoplasia of the left transverse sinus, specifically when the SSS drains only to the right transverse sinus (variation type 2), the left transverse sinus was found to be hypoplastic. Also, when the SSS drained to both the right and left transverse sinuses, but the SS only drained to the right transverse sinus (variation type 8), the left transverse sinus was hypoplastic.

Furthermore, variation type 2 was significantly associated with no hypoplasia of the right transverse sinus. However, hypoplasia of the right transverse sinus was significantly associated with variation type 3 when the SSS drained only to the left transverse sinus. Hypoplasia of the right transverse sinus was also significantly associated with variation type 4, when the SSS bifurcated and drained to the left and right transverse sinuses respectively without draining into the confluence of sinuses. This might be indicative of an unequal distribution of blood flow from the SSS to the transverse sinuses in cases where the SSS bifurcates and does not drain into a confluence.

Thus, variations in the drainage pattern of the SSS, dominance and hypoplasia of the transverse sinuses are closely associated and these should all be taken into account when planning any cerebral interventions.

5.6 Accessory vessels

Analyses of this variable proved challenging due to the low number of cases in which the accessory vessels were absent. This was especially true for the vein of Galen, which was present in 299 of 302 cases and the intracerebral veins, present in 300 of 302 cases. The left vein of Trolard was most frequently absent (6 out of 302), but it was also the most difficult vessel to identify accurately on imaging whereby 128 cases of 302 were unknown.

The veins of Labbé were more easily identified than the veins of Trolard. This could be due to the vein of Trolard tending to be smaller in diameter than the vein of Labbé and also due to the location of the vein of Trolard because it is obscured behind the gyri of the parietal lobe.

There is no significant difference in the presence of accessory vessels between males and females, which implies that their presence is not affected by sex chromosomes or hormonal difference.

It is extremely unlikely that an accessory vessel will appear or disappear after birth therefore age and presence/absence of accessory vessels were not analysed.

The presence or absence of accessory vessels were not associated with the type of circulatory dominance but did have an influence on the actual venous sinus volume as seen in section 5.3.5.

5.7 Variations of the cerebral venous system

5.7.1 Variations of the drainage pattern of the superior sagittal sinus

The SSS is formed from the marginal sinus of each side during embryonic development. The marginal sinus becomes approximated towards the midline as it forms the superior sagittal plexus (Knapp & Schmideck, 1984). The SSS may remain separated into two limbs (posteriorly not fusing at the midline), each of which drain laterally into their respective transverse sinuses on each side. The two limbs usually join to form the confluence of sinuses during the sixth fetal month (Widjaja & Griffiths, 2004).

The variation type 4 of the SSS is the result of non-fusion of the two limbs of the SSS in the midline.

The types of variation for the current study differ slightly from the types described by Bisaria in 1985. Bisaria described type 1 as being where the SSS drains into one lateral sinus and the SS into the other lateral sinus, with no connection between the two. The current study expanded on this by differentiating left and right, thus type 5 and 6. Bisaria (1985) noted type 2 as cases in which the SSS and the SS fork, and the forks from both sinuses join to form the lateral sinuses, this correlates to variation type 4 in the current study.

Kitamura *et al.* (2017) describe the scenario where the SSS drains directly into one of the transverse sinuses leaving the other transverse sinus hypoplastic or absent; for the current study this corresponds to variation type 2 and 3.

In the current study it was found that 37.09% of individuals had a type 1 classification of the SSS, thus the most common configuration. The remainder of the individuals displayed types 2, 3 and 4 (62.91%) with the most common variations present were type 2 (25.83%), type 4 (17.88%) and type 3 (11.26%) in order of decreasing frequency.

No significant association was found between variation of the SSS and sex (p=0.489), age groups (p=0.343) or presence of pathology (p=0.491).

The study found a significant association between variation of the SSS and the type of circulatory dominance (p<0.001). Variation type 2 appears to be associated with right sided dominance and variation type 4 appears to be associated with co-dominance. This is in line with the findings of increased volume in the right sigmoid sinus in conjunction with type 2 variation. The other types on the drainage pattern of the SSS did not show a significant association with dominance.

5.7.2 Variations of the confluence of sinuses

One should consider the confluence of sinuses as a complex system of anastomosing sinuses and not just the drainage point of the SSS (Durgun *et al.,* 1993).

The terminal or jugular portion of the transverse sinus develops first. At the point of entry of the superior petrosal sinus to the jugular fossa (the sigmoid portion), the transverse sinus consists of a single large channel and has the same tributaries and relations as in the adult. The proximal part of the transverse sinus is not as well established as the terminal part. The large capillary meshwork along the dorsal margin shows that the blood channels here are still in their formative stages and represents the remainder of the anterior dural plexus. Along the anterior margin of this plexus the main channel is forming, into which the inferior cerebral vein empties. In adjusting to the growth of the hemispheres this channel is seen migrating posteriorly and assuming a more horizontal course. The change in direction along with the growth in the main channel (at the expense of the formative meshwork) remains to be completed before the adult configuration can be considered established. Variations in the region of the confluence sinuum in the adult can be understood as variations in the channel selection through this tentorial meshwork (Streeter, 1915).

In the current study more than half of the individuals (58.28%) had a variation in the pattern of the confluence or no confluence at all and only 41.72% of persons had a true confluence. This is indicative of the fact that the confluence of sinuses is variable in its configuration and might be susceptible to different influences during embryological development.

No significant association was found between sex, age groups or presence of pathology and variations of the confluence of sinuses. This leads one to think that the configuration of the confluence of sinuses does not change after birth.

Of note is that variations of the confluence of sinuses showed no significant association with cerebral circulatory dominance. Thus, it is thought that the drainage from the SSS into the transverse of sinuses influences dominance, with or without the confluence of sinuses.

5.7.3 Other variations of the dural venous sinuses

For the current study variations of the dural venous sinuses were gouped into 20 types. The first eight types cover variations of the transverse sinuses.

The transverse sinuses will continue to enlarge up until the sixth fetal month, afterwhich they there size stabilises. There is a transition period from the plexiform network stage to that of rapid increase and then a decrease in size, during which time the dural sinuses in the region of the confluence of sinuses may become irregular shaped (Okudera *et al.,* 1994).

The most common variation seen in the right dominant transverse sinus, followed by co-dominance and then left dominance. The transverse sinus was found to be absent in two patients on the right and in three patients on the left. No cases of multiple transverse sinuses were noted. This shows that although dominance might be variable between individuals, the transverse sinus is consistantly present and is not duplicated. This indicates that the plexiform network disappears during development and dominance may be attributed to the increase and decrease in size of the transverse sinuses.

Variation in the drainage pattern of the SSS is seen in variation types 9, 10, 12 and 13. Multiple SSS were not seen in any cases. The SSS was seen to bifurcate and then drain into both the left and right transverse sinuses in 64 cases and draining only to one transverse sinus in 113 cases. It is thought that when the marginal sinus of each side does not fuse in the midline to form the SSS, the SSS is seen to have two "legs" or bifurcating. The SSS drained to one transverse sinus and the SS to the other transverse sinus in 13 cases. Various influences during gestation may be the cause of the SSS and SS draining to different transverse sinuses and not into the confluence of sinus as generally described.

The SS bifurcated and drained to both the left and right transverse sinuses in 112 cases. The SS was not absent in any cases and it was also not duplicated in any cases. The SS forms part of the deep venous system of the brain, the deep system is much more constant in its appearance than the superficial systems. It is thought that the SS is thus not really influence by various factors during embryological development. The inferior sagittal sinus was absent in only two cases and not duplicated in any cases in the current study. The same is true for the inferior sagittal sinus, as it is part of the deep system it might have an inherent protection against developmental changes and variations.

The sigmoid sinus was absent on the left in two cases and always present on the right. In both cases where the sigmoid sinus was absent on the left, the transverse sinus on the left was extremely small and hypoplastic. It is thought by the researcher that the blood from the left transverse sinus drained through an alternative route via the petrosal sinuses.

The variation in the pattern of the confluence of sinuses were seen in 78 patients, which mostly corresponded to a bifurcating SSS and/or SS. The variations of the confluence are described in section 5.7.2 above.

5.8 Influence of pathology on the morphology of venous sinuses

In the current study the presence or absence of a pathology was known for 290 patients. Of these cases, the majority (245; 84.48%) exhibited some form of pathology. It should be noted that as the sample came from two hospitals the sample does contain a bias as persons presenting to hospital for neurological imaging would be expected to have some form of pathology.

A significant association was found between sex and the presence of pathology. It was determined that males are 2.3 times more likely to present with pathology than females. The sample consisted of more females than males, however the males were more prone to present with pathology than females. It should be noted that certain diseases (e.g. lifestyle, certain cancers) (Hattori *et al.,* 2017) present more in one sex than in the other. Even though the presence of pathology was noted for the sample, the type of pathology was not, as complete medical notes and diagnoses for all individuals were not available.

No association was found between age groups and the presence of pathology. However, when the median age between persons with pathology was compared with persons without pathology a p value of 0.051 was found, this is almost significant. The median age of persons without pathology was slightly lower, 42 years, than those with pathology, 48 years. One could argue that as a person ages the presence of pathology is more common (Niccoli & Partridge, 2012; Jaul & Barron, 2017). The median age was therefore greater in persons without pathology than those without, however the study is focused on presence of cerebral pathology only which might not be age dependent.
The current study found no significant association between circulatory dominance, variations of the SSS or variations of the confluence of sinuses and the presence of pathology. Thus, having a variation of the SSS, confluence of sinuses or being dominant to one side does not predisposed an individual to the formation of pathology.

5.9 Venous blood pools

The major dural venous sinuses will only reach their final configuration some months after birth. Early fetal dural venous connections are plexiform in nature and are constantly changing Mack *et al.,* 2009). The primitive dural plexus is continually adjusting due to the dramatic growth of the cerebral hemispheres in utero. These adjustments are largely possible due to spontaneous migration of the principle dural veins, and for this a venous plexus is essential (Mack *et al.,* 2009).

No venous blood pools were found for any of the patients. A sample of fetuses and children might show venous blood pools present, because the current sample consisted of adolescence, adult and elderly the blood pools may all have regressed. A younger sample of children and fetuses might show venous blood pools still present.

5.10 Venous sinus configuration as a method of identification

Although the I-Number assigned to each person was not found to be unique to that individual it does cast some light on the venous patterns that tend to repeat within a sample.

The I-Number is unique enough to determine the likely identification of an individual. An example could be if a MR scan exists for a missing person one can determine the I-Number and rule out many individuals who do not match.

The I-Number may also be of value in the surgical setting as it gives a high-level overview of the entire venous drainage pattern of the individual.

5.11 Strengths and limitations

5.11.1 Practical implications

The current study illuminated on the implications of cerebral venous drainage variations on accurately interpreting cerebral vascular imaging and following the correct surgical pathway to limit mortality and morbidity.

With limited information available on imaging of the intracranial veins and sinuses, especially in young adults and children, it is important to have an understanding of the normal anatomy of the cerebral venous system and its anatomical variations to provide background to any future studies on anomalous venous structures.

The study provided a novel way of interpreting dominance of the cerebral venous system and well as understanding the implications thereof.

Reports in the literature showed a lack of comprehensive classification with regards to variations within the confluence of sinuses and the SSS (Bayarogullari *et al.*, 2018). The current study provided a novel and clear classification of variations of the confluence of sinuses as well as the drainage pattern of the SSS; which led to a better understanding of the influence of variation on venous drainage dominance within individuals.

Furthermore, the current study created a tracing protocol to enable physicians, specialist and radiologist to determine cerebral venous dominance and trace the borders of the cerebral venous sinuses. Such a protocol ensures standardisation of venous volume measurements and accurately determines venous blood flow through the venous sinuses.

The current study created a novel approach in using cerebral vasculature as an identification method, for use when an individual cannot be identified by normal techniques (fingerprints and/or DNA). This novel approach may also be applied to obtain a complete overview of the venous drainage pattern of individuals when surgical intervention is considered.

5.11.2 Limitations of the study

A limitation of the study is that it was conducted in only two centres, which may cause a selection bias in the sample and some variations frequencies may change. Furthermore, as the persons presenting to the centres are already thought to have a cerebral pathology present, this may cause a bias when comparing cerebral venous variation with the presence or absence of cerebral pathology.

The lack of long-term follow-up data to look for significance of anatomical variations found in the population and their association with the development and/or treatment of cerebral pathologies later in life. As this was a cross sectional study, it is difficult to show the impact of age on the cerebral venous system, as one would with a longitudinal study.

The current study did not measure the occipital sinus as its margins were difficult to visualise om MR imaging. Further studies could build on the studies by Widjaja & Griffiths (2004) who tried to determine if there is a correlation between the presence of the occipital sinus and the age and transverse sinuses of individuals.

It was found that insufficient variables were used in the determination of the identification number (I-Number) of patients. Future studies could build on the current research by including more variables in calculating the I-Number.

5.11.3 Future research

The current study has laid the foundation for future research involving comparing growth factors gene expression with gross anatomy and variations. It is yet to be determined how the expression of certain gene influence gross anatomical variation of the cerebral venous sinuses.

Information regarding variations of the dural venous sinuses using contrast enhanced volume tracings is lacking for other countries and population groups outside of South Africa, therefore comparisons were not possible. Future studies are needed to compare the volumes of dural venous sinuses between different populations and geographical areas.

A novel idea would be to use cerebral vessel configuration and drainage patterns as a method of identification. The hypothesis being that venous vessel volume, configuration and variation is unique to each person, due to genetics, environment and pathology. If one were to combine venous sinus volume with arterial configuration one would establish a pattern unique to an individual which could be used for identification in the absence of fingerprints and/or DNA. This unique number will also be of value when a high-level overview of the venous drainage pattern of an individual is needed.

5.11.4 Reflections

The fetal dissections contributed to a complete view of the dural venous sinus in a real life setting without influence of software or filters. Increasing the number of fetuses to be dissected will contribute to a better understanding of the fetal drainage pattern, especially as MR imaging of young children is scarce.

The exact mechanisms influencing variations, such as hormones and genetics, were outside the scope of this study. However, it would be of value to research these aspects further and combine them with the results and variation patterns seen in the current study.

Initially the researcher obtained only one measurement of the diameter of each sinus for the statistical analyses and dominance determination. However, the literature could not agree where this measurement should be made. Thus, after consultation with a neuroscientist at the University of Cape Town it was decided to trace the entire volume of the venous sinus and use the volume for statistical analyses and dominance determination. By using the venous sinus volume for morphological and statistical analyses a clear and descriptive model of the venous sinuses could be made. The three-dimensional models enabled the researcher to view the venous sinuses in a new way as complete structures.

Manually sorting through the MR images available at the Groote Schuur and Red Cross Children's War Memorial hospitals and then tracing each venous sinus proved to be a very labour intensive and time-consuming process. However, the three-dimensional models that were constructed proved invaluable.

The free software Horos was used to trace the venous sinuses. The software proved user friendly and more than adequate for research. The software automatically created the three-dimensional modules needed and also calculated the volume of each structure. Tracings could be saved and reviewed again at a later date.

Although many studies and classifications have been made regarding the variations of these structures, there is a lack of a comprehensive classification that includes all variations. The goal of the current study was to determine the anatomical variations more comprehensively particularly at the level of the SSS and confluence of sinuses. The study improved on the current literature by using contrast enhanced images as opposed to non-enhanced images or results obtained at autopsy. It is also the first study to establish a tracing protocol for venous volume to determine cerebral dominance and describe variations of the dural venous sinuses.

Chapter 6: Conclusions

To date, a comprehensive description of the cerebral venous drainage structures and their variations have not been found for a South African sample. This research project is the first to describe the cerebral venous drainage patterns for adolescents (13 – 17 years), adulthood (18 – 64 years) and the elderly (65 years and older) for such a South African sample. Furthermore, the study described the embryological and fetal development of venous drainage patterns by means of a comprehensive literature review and a limited number of fetal dissections. The study speculated on the possible developmental mechanisms behind variations observed in the cerebral venograms and discussed the clinical impact of cerebral venous drainage patterns in imaging and surgical interventions and their potential use in the identification of persons.

The study included MRIs from 302 individuals obtained from the Groote Schuur and Red Cross Children's War Memorial Hospitals. The study described the macroscopic anatomical venous drainage patterns using cerebral venograms for adolescents, adults and the elderly.

The study computed three-dimensional models of the dural venous sinus to calculate the volumes of these sinuses and macroscopically compare the venous drainage patterns. Dural venous sinus volume was determined for the following variables: sex, age, presence of pathology, presence of accessory vessels, dominance of venous drainage and hypoplasia, as well as between the left and right sides of all individuals.

It was found that the sigmoid and transverse sinuses were significantly larger on the right then the left. Males were shown to have significantly larger sinus volumes compared to females for all the sinuses except the left sigmoid sinus.

In the current study the volume of the SSS was compared between age groups and it was found that the volume of the SSS was significantly decreased in the older age groups. There was no significant association between presence of pathology and venous sinus volume.

The volumes of the superior sagittal and right sigmoid sinuses were found to be smaller when the left vein of Trolard was absent. Furthermore, when there was no confluence of sinuses present the volume of the right sigmoid sinus was significantly greater than in cases where a confluence of sinuses was present.

Variations of the drainage pattern of the SSS were grouped into nine types. It was seen that these variations of the SSS did not significantly influence the actual volume of the SSS. However, variations of the SSS did significantly influence the volumes of the right and left transverse and the right and left sigmoid sinuses.

Circulatory dominance was visualised as a continuous spectrum and the study was the first to use venous sinus volume in the determination of cerebral circulatory dominance. Circulatory dominance was seen in the sample as follows; right dominance (41.72%), co-dominance (39.4%) and left dominance (18.87%). No significant difference was seen for cerebral dominance between males and females or between persons with or without pathology.

A significant association was found for individuals where the SSS bifurcated into the right and left transverse sinuses and circulatory co-dominance. A significant association was also seen between right cerebral circulatory dominance and the SSS draining only to the right transverse sinus.

The right transverse sinus was found to be hypoplastic in 234 individuals and the left transverse was hypoplastic in 266 individuals. Females also showed significantly more hypoplasia of the right transverse sinus than did males. No association was seen between sex and hypoplasia of the left transverse sinus.

Circulatory dominance and hypoplasia did show a close relationship, as right dominance was significantly more common in persons with hypoplasia of the left transverse sinus. Left dominance was significantly more common in persons with hypoplasia of the right transverse sinus.

The generally acceptable configuration of the SSS (the SSS drains completely into the confluence of sinuses) was found in 37.09% of individuals. The SSS drained only to the right transverse sinus in 25.83% of individuals. The SSS bifurcated and drained to the right and left transverse sinuses in 17.88% of individuals. The SSS drained only to the left transverse sinus in 11.26% of individuals.

A significant association was also found between variations of the drainage pattern of the SSS and hypoplasia of the transverse sinuses. In cases where the SSS drained only to the right transverse sinus, the left transverse sinus was found to be hypoplastic and the right transverse sinus was not associated with hypoplasia. Furthermore, in cases where the SSS drained to both the right and left transverse sinuses, but the SS only drained to the right, no hypoplasia was seen on the left. In cases where the SSS bifurcated and drained to the transverse sinus without entering the confluence of sinuses first, the right transverse sinus was more frequently hypoplastic than the left. More individuals showed hypoplasia of the right transverse sinus when the SSS drained only to the left transverse sinus.

In the current study 41.72% of individuals had a true confluence of sinuses, 32.45% had no confluence and 25.83% had variation of the confluence of sinuses.

The presence of pathology was known for 290 individuals in the sample and of these 245 (84.48%) had some form of visible cerebral pathology.

The internal cerebral veins were most easily identified and were present bilaterally in 300 individuals, the vein of Galen in 299, the right vein of Labbé in 239, left vein of Labbé in 236, right vein of Trolard in 181 and the left vein of Trolard in168 individuals.

The study showed that the cerebral venous drainage patterns are unique to each person and a more individualised approach may be of value in treating patients with cerebral vascular diseases.

Even though no statistically significant association was found between the presence of pathology and variation in the cerebral venous drainage patterns, the drainage pattern does impact on the vascular treatment of cerebral pathology.

The study considered all factors in compiling a complete description of the venous drainage patterns seen in the South African sample. A novel approach to determining cerebral circulatory dominance was taken and it is the first study to compile three-dimensional structures of the dural venous sinuses which may be used in descriptions and analyses of variations the venous sinuses.

Based on the relevant literature the results of the study suggest that there is a possibility of the persistence of embryological venous drainage patterns in some individuals.

The study built on previous literature by improving the methodology and techniques for studying and describing the dural venous structures. The approach in the study was not to use cadavers or unenhanced images, but rather to include state of the art tracing software and contrast enhanced images to provide a detailed description of venous drainage patterns and their variations.

Finally, the study proposed a new technique to obtain a high-level overview of the cerebral venous drainage patterns by assigning an identification number for each patient. This I-Number may also be used in the future for identification purposes showing the contribution clinical anatomy may have in the field of forensic sciences.

Appendix A: University of Cape Town ethics approval

UNIVERSITY OF CAPE TOWN
Faculty of Health Sciences

Human Research Ethics Committee

Room E53-46 Old Main Building
Groote Schuur Hospital
Observatory 7925
Telephone [021] 404 7682
Email: nosi.tsama@uct.ac.za
Website: www.health.uct.ac.za/fhs/research/humanethics/forms

07 September 2016

HREC REF: 483/2016

Prof G Louw
Human Biology
Anatomy Building

Dear Prof Louw

PROJECT TITLE: INCREASED RISK OF CEREBRAL VASCULAR ACCIDENTS DUE TO PERSISTENCE OF EMBRYONIC VENOUS PATHWAYS IN THE ADULT BRAIN (PhD Candidate – Dr F du Toit)

Thank you for submitting your study to the Faculty of Health Sciences Human Research Ethics Committee for review.

Before approval can be given, the investigator must address the following issue/s raised by the Committee:

- Please clarify whether MRI's or venograms of patients currently receiving care will be used in this study; or whether this study will review MRI's or venograms of patients retrospectively. If images of patients currently receiving care will be included in this study, then informed consent and assent will be required. In addition, Form A for non-therapeutic research involving minors will need to be completed (this form is available on the HREC website).
- Under ethical considerations, please state the risks and benefits of this study.
- Under ethical considerations, please refer to the 2015 version of the Department of Health, Ethics in Research guidelines.

We acknowledge that the student Dr F du Toit will be involved in this study.

Please note that no research may occur without formal written HREC approval.

Please quote the HREC reference number in all your correspondence.

Yours sincerely

T Burgess

pp

PROFESSOR M BLOCKMAN
CHAIRPERSON, FHS HUMAN RESEARCH ETHICS COMMITTEE

Appendix B: Approval from Groote Schuur Hospital and the Department of Health of the Western Cape

GROOTE SCHUUR HOSPITAL
Enquiries: Dr Bernadette Eick
E-mail : Bernadette.Eick@westerncape.gov.za

Professor G. Louw
Human Biology
Anatomy Building

E-mail: francescadutoit@gmail.com

Dear Professor Louw

RESEARCH PROJECT: Increased Risk of Cerebral Vascular Accidents Due to Persistence of Embryonic Venous Pathways in the Adult Brain (PhD Candidate Dr F. du Toit)

Your recent letter to the hospital refers.

You are hereby granted permission to proceed with your research which is valid until **30 October 2017.**

Please note the following:

a) Your research may not interfere with normal patient care.
b) Hospital staff may not be asked to assist with the research.
c) No additional costs to the hospital should be incurred i.e. Lab, consumables or stationary may be used.
d) **No patient folders may be removed from the premises or be inaccessible.**
e) Please discuss the study with the HOD before commencing.
f) Please introduce yourself to the person in charge of an area before commencing.
g) Please provide the research assistant/field worker with a copy of this letter as verification of approval.
h) Confidentiality must be maintained at all times.
i) Should you at any time require photographs of your subjects, please obtain the necessary indemnity forms from our Public Relations Office (E45 OMB or ext. 2187/2188).
j) Should you require additional research time beyond the stipulated expiry date, please apply for an extension.
k) **On completion of research, please submit a copy of the publication or report.**

I would like to wish you every success with the project.

Yours sincerely

DR BERNADETTE EICK
CHIEF OPERATIONAL OFFICER
Date: 28 October 2016
BE/vms

C.C.   Mr L. Naidoo, Professor E. Weimann, Professor S. Beningfield, Mrs N. Behardien-Peters
G46 Management Suite, Old Main Building,   Private Bag X,
Observatory 7925   Observatory, 7935

Tel: +27 21 404 6288   fax: +27 21 404 6125   www.capegateway.go.v.za

Appendix C: Venous sinus configuration as a method of identification

Patient Number	I-Number	Patient Number	I-Number	Patient Number	I-Number
001	111110	031	213110	061	211230
002	231110	032	213210	062	213110
003	222310	033	1x1114	063	211230
004	222410	034	131230	064	211230
005	213110	035	222420	065	211110
006	113310	036	221110	066	231210
007	111210	037	103420	067	213210
008	131110	038	211411	068	223830
009	113420	039	233310	069	131530
010	113420	040	222110	070	231230
011	231210	041	213110	071	211110
012	212110	042	222110	072	131230
013	213310	043	103120	073	113530
014	111210	044	211110	074	111530
015	213420	045	211210	075	111230
016	211420	046	122120	076	211230
017	111420	047	111210	077	231230
018	213110	048	212420	078	122110
019	211210	049	2x1114	079	112330
020	213120	050	213110	080	131110
021	131230	051	211120	081	231210
022	232114	052	131120	082	113330
023	233110	053	212110	083	2xx710
024	131210	054	213420	084	112830
025	211210	055	212420	085	213210
026	113210	056	131510	086	133420
027	212310	057	213420	087	122830
028	111220	058	232114	088	131111
029	211420	059	111420	089	213110
030	213110	060	113230	090	213420

Patient Number	I-Number	Patient Number	I-Number	Patient Number	I-Number
091	213420	121	231230	151	211110
092	231110	122	211230	152	1x1110
093	111110	123	211230	153	213630
094	211510	124	203110	154	231110
095	112110	125	211530	155	211230
096	213120	126	111220	156	222110
097	212720	127	213420	157	222110
098	131230	128	211110	158	131420
099	111710	129	111230	159	131110
100	231230	130	212110	160	213230
101	213110	131	131220	161	113420
102	211230	132	211230	162	212420
103	222330	133	213230	163	113110
104	212330	134	113110	164	213330
105	212330	135	113110	165	122110
106	131230	136	213420	166	211110
107	211110	137	112320	167	111230
108	113420	138	211420	168	213420
109	213730	139	213420	169	231230
110	221234	140	212420	170	113330
111	131330	141	113110	171	111420
112	231110	142	113110	172	213110
113	212420	143	213330	173	213230
114	122230	144	231230	174	211110
115	113110	145	211230	175	211230
116	113420	146	113110	176	211230
117	213110	147	113420	177	231420
118	213110	148	113230	178	222420
119	113330	149	223730	179	122330
120	131110	150	113420	180	112420

Patient Number	I-Number	Patient Number	I-Number	Patient Number	I-Number
181	211230	211	213110	241	213110
182	131220	212	212330	242	113425
183	211230	213	123110	243	113110
184	213420	214	213310	244	113230
185	211110	215	133120	245	231530
186	211110	216	213330	246	213110
187	231230	217	131230	247	111230
188	213110	218	112330	248	111230
189	213420	219	212110	249	113530
190	103230	220	111820	250	212425
191	113230	221	212330	251	211230
192	213320	222	112420	252	212330
193	231230	223	121930	253	213110
194	213110	224	113330	254	131420
195	113230	225	213110	255	213420
196	211230	226	113230	256	231230
197	103630	227	231120	257	212420
198	112110	228	113410	258	211230
199	231220	229	222110	259	231230
200	213310	230	213420	260	213330
201	112320	231	233110	261	213110
202	112420	232	131110	262	111110
203	211110	233	213620	263	111110
204	113120	234	113420	264	113110
205	212110	235	212310	265	122420
206	212110	236	113210	266	231110
207	213120	237	113120	267	211420
208	131120	238	131110	268	211120
209	213630	239	111110	269	113110
210	211230	240	233110	270	212230

Patient Number	I-Number	Patient Number	I-Number
271	211110	301	111110
272	133120	302	213110
273	112330		
274	223420		
275	223110		
276	213230		
277	113530		
278	111420		
279	213330		
280	211230		
281	131110		
282	213230		
283	113110		
284	211110		
285	131230		
286	212120		
287	111110		
288	122420		
289	112730		
290	223330		
291	213110		
292	122110		
293	212110		
294	131110		
295	211110		
296	211420		
297	213330		
298	213230		
299	213330		
300	131230		

www.ingramcontent.com/pod-product-compliance
Lightning Source LLC
LaVergne TN
LVHW042109190726
843493LV00006B/1413